PHILOSOPHY
OF
BIOLOGY
TODAY

SUNY Series in Philosophy and Biology
David Edward Shaner, Furman University, editor

PHILOSOPHY
OF
BIOLOGY
TODAY

Michael
Ruse

STATE UNIVERSITY OF NEW YORK PRESS

Published by
State University of New York Press, Albany

For information, address State University of New York
Press, State University Plaza, Albany, N.Y., 12246

Library of Congress Cataloging in Publication Data

Ruse, Michael.
Philosophy of biology today: Michael Ruse. p.cm. — (SUNY series in
        philosophy and biology)
Bibliography: p.
Includes index.
ISBN 0-88706-910-X. ISBN 0-88706-911-8 (pbk.)
1. Biology — Philosophy.   I. Title.   II. Series.
QH331.R8793   1988   88-15377
574'.01 — dc19   CIP

10 9 8 7 6 5 4 3 2

For Marjorie Grene and Ernst Mayr,
with thanks.

Contents

Preface

This small volume is a handbook to the philosophy of biology. As such, I hope it will introduce newcomers to the field and prove useful as a source of references to those who are already working on problems in the subject. It is not particularly inspired or inspiring. However, it is the kind of thing I would have liked someone else to have done for me. As soon as I began writing the text, I realized that I could not possibly avoid showing some of my own interests — call them prejudices if you will — so at once I stopped worrying. Nor shall I apologize, since I know that my good friends in the field will delight in warning their students of my biases. Nevertheless, having written an indulgent first draft, which I rather liked, I have now toned down the text, somewhat.

A much shortened version of this volume served as the base of the first Gordon Lowther memorial address at York University, Canada, in February 1988. I am happy to acknowledge this fact and to thank Jane and Robert Haynes for deeming me worthy to give such an address. A word about my dedication is also in order. I have worked as a philosopher of biology for twenty years now. I think if I were asked to name two people with whom I have had the strongest philosophical differences, the philosopher Marjorie Grene and the biologist Ernst Mayr would be at the top of the list. This means, as is the nature of things, that they are the two from whom I have learnt most. Like everyone who has touched on the philosophy of biology, I am incredibly in their debt, most particularly for showing me that the philosophy of science does not necessarily have to be the philosophy of physics.

I am a lucky person. My work is my hobby. I cannot imagine why anyone would not want to spend their life doing philosophy. It is the most important subject that there is. It is also great fun, particularly when you are surrounded by a group of supportive fellow workers. I am happy to acknowledge the following people in particular for the help that they gave on the text and/or bibliography: Francisco Ayala, John Beatty, William Bechtel, Michael Bradie, Robert Brandon, Antonio Britto da Cunha, Daniel Brooks, Donald Campbell, Arthur Caplan, Camilo Cela-Conde, Lindley Darden, Richard Dawkins, Michael Ghiselin, Marjorie Grene, James Griesemer, David Hull, Philip Kitcher, Scott Kleiner, James Lennox, Rolf Lother, Steven Marshall, Ernst Mayr, Peter Richerson, Alexander Rosenberg, David Shaner, Elliott Sober, Neil Tennant, Paul Thompson, Wim van der Steen, George Williams, Edward O. Wilson.

Introduction

S cience aims to understand the physical world about us. Even its critics concede that, in this century particularly, it has been highly successful in its aims. It is no surprise, therefore, that philosophy — which aims toward understanding the nature of knowledge itself — has been so much interested in the achievements of science. Nor is it a surprise that, in this century particularly, the philosophy of science has become almost a subdiscipline in itself.

But this does not include the philosophy of biology — at least, it did not until very recently. Although the father of the philosophy of science, Aristotle, was as much a biologist as a philosopher, the philosophers of science in the twentieth century have focused mainly on the physical sciences, and any spare effort has tended to be directed toward the social sciences. What little attention has been paid to biology has been generally directed to one extreme or another. At one end of the spectrum we have those who were overly impressed by the turn-of-the-century formalisms of the logicians and mathematicians, and who wanted to do likewise for biology. Since they — especially their leader J. H. Woodger (1937, 1939, 1952, 1959) — were simultaneously empiricists of the most naively dogmatic kind, their efforts tended to go unread (Smart, 1953; Ruse, 1975a; although, see recently, Rizzotti and Zanardo, 1986, and Zanardo and Rizzotti, 1986). At the other end of the spectrum we have those who feared and loathed materialism, and who were determined to prove that an understanding of organisms demands reference to vital forces or spirits — elans vitaux or entelechies — forever beyond the grasp of conventional science (Bergson, 1913; Driesch, 1905, 1914). Unfortunately, these proved no more useful to our understanding than did the unseen forces against which Francis Bacon raged three centuries ago.

Things have changed. Modern biology has become one of the most vibrant of empirical inquiries. Molecular theories and techniques have transformed genetics. Evolutionary theory has had its most exciting two decades since Darwin. Human beings have been brought into the fold, by specialists ranging in disciplines from paleoanthropology to sociobiology. Systematics has thrown up new theories and techniques. And social issues abound, from the propriety of recombinant DNA research to the fight against Creationism, a movement whose members want the book of Genesis given equal billing with evolutionary theory by public school teachers of the life sciences.

With progress and controversy come conceptual challenges. We now have an increasing number of people from both biology and philosophy, who look at the nature and status of biological models and theories themselves. And there are also those, with growing voice, who argue conversely, that philosophy can profit from an appreciation of modern thought about the world of organisms. It really does matter to the traditional problems of epistemology (theory of knowledge) and ethics (theory of morality) that we are modified monkeys, not the special creations of an all-loving God some six thousand years ago.

In the firm belief that both biology and philosophy benefit from their mutual interaction — and as one whose belief has led him to start the new journal *Biology and Philosophy* — I now offer a rapid survey of the growing field of the philosophy of biology. (For older surveys of the state of the art, see Beckner, 1959; Hull, 1969, 1974a; Ruse, 1973a; Munson, 1971; Ayala and Dobzhansky, 1974; and Waddington, 1968, 1969, 1970. More recent work includes Hull, 1979b, 1982b; Grene, 1986; Sattler, 1986; Rosenberg, 1985a; and Ruse, 1988 a, b, c — although see Stent, 1986; and Rosenberg, 1986b. See also Gotthelf and Lennox, 1987, for the philosophy of biology of Aristotle.)

CHAPTER 1

EVOLUTIONARY THEORY

A choice of topics already imposes values on one's subject. Indeed, even the order of topics is beyond neutrality — should one start any treatment of biology at the molecular end, or is this kowtowing to the physicists? Believing strongly that one cannot understand a subject without knowing its history, I will begin with evolutionary theory taken as a whole. If justification be needed for my choice of *evolutionary* biology, I would remind you of the oft-made remark of the great geneticist Theodosius Dobzhansky (1973): "Nothing in biology makes sense except in the light of evolution."

No one can deny that the publication in 1859 of Charles Darwin's *Origin of Species*, in which he argued that evolution is a function of natural selection brought about by a struggle for existence, was a landmark. On that date, biology got what the historian and philosopher Thomas Kuhn (1962) has described as a "paradigm." I hasten to add that although Darwin's theory undoubtedly has paradigmatic status — if it does not, what would have? — one should not go on to conclude that Kuhn's stimulating thesis about the general nature of scientic change applies to the great biological revolution of the nineteenth century. In fact, as several people have shown, it almost certainly does not (Mayr, 1971; Ruse, 1970; Greene, 1971). Kuhn sees abrupt, discontinuous changes, from one way of thinking to another, much as occurs in a gestalt switch. But, whether this be generally true, there is no question at all but that Darwin's revolution was continuous. He was highly indebted to his nonevolutionary predecessors, adopting many of their ideas and ways of thinking, and adapting them to his own evolutionary ends (Ospovat, 1981; Ruse, 1979a; Hull, 1985b). For instance, he drew heavily on results in embryology, paleontology, anatomy, biogeography, and more. His genius was to show how the already-known facts fit into the evolutionary picture. Typical was Darwin's seizing on von Baer's discussion of embryology — where it was shown that organisms of different species frequently have very similar embryos — and arguing that such a picture makes perfect sense to the evolutionist. Similarity of embryo is a mark of shared ancestry.

Most fascinating and pertinent to us today is the now-realized extent to which Darwin drew on the ideas of the philosophers of his day. In particular, he agreed entirely with the authorities (especially with the philosopher-scientists John F.W. Herschel and William Whewell) that the ideal of a scientific theory is Newtonian mechanics and that his own theory ought to emulate it (Ghiselin, 1969; Ruse, 1975b). I should say, however, that even then there were those who queried just how successful Darwin had been — or, indeed, how successful any evolutionist could be (Hull, 1973a). There were complaints that Darwin's theory was much looser than the best theories of the physical sciences, and there were doubts about whether any biological theory could do better. (Recent philosophical discussions of Darwin's work and its consequences include Beatty, 1982b; Hodge, 1977, 1983, 1988;

Kleiner, 1985, 1988; Lloyd, 1983; Oldroyd, 1986; Provine, 1983; Recker, 1987; Rosenberg, 1985c; Waters, 1986b; Williams, 1986; Kitcher, 1985b; and many contributions to Kohn, 1985.)

These are worries which persist to this day; so rather than lingering in the past, let us at once bring the inquiry up to the present. Today's dominant evolutionary theory is a synthesis of Darwin's ideas and modern genetics (Beatty, 1986). On the one hand, we have the Darwinian insights about selection, and how they give rise to adaptation. On the other hand, we now have a full theory of the nature of heredity and of how variations are produced and transmitted from generation to generation, an area about which Darwin himself was almost entirely ignorant (or mistaken). There are queries about and rivals to this "neo-Darwinian" picture, and we shall be looking later at some of the more important. For the moment, however, the dominant view can serve as background, and at once we plunge into what are perhaps *the* philosophical questions about biology, especially evolutionary biology.

Recognizing that physics itself has changed in the past century, I would ask: Is evolutionary theory a science of the same kind as one finds in physics and chemistry? Does the biologist think like the physicist? Does the biologist produce results like the physicist? Giving the query some real content, let me ask three specific questions. Is evolutionary theory (neo-Darwinism) an axiomatic theory of the kind that many believe physicochemical theories aspire to? Is neo-Darwinism a genuinely empirical theory, or is it just a metaphorical redescription of nature, more akin to poetry than to science? And does neo-Darwinism unify our ideas and thoughts, as does the best science? (Nagel, 1961, and Hempel, 1966, are the *loci classici* of the traditional physics-based view of science. That not all would accept this view, even for physics, will be noted later.)

Taking the first question, even its best friends must concede that neo-Darwinism is looser (unkind critics would say "flabbier") than the best physics (Scriven, 1959; Smart, 1963). There is frequently no tight connection between premises and conclusions. As a consequence, we simply do not seem to have the predictive power in evolutionary studies that we can demand in physics.

An astronomer can tell us just when an eclipse of the sun is expected. No biologist would dare predict the future history of the elephant's trunk. The premises simply do not bear a conclusion — about particulars or about generalities — one way or the other.

Several separate options now present themselves, and philosophers have divided all ways. First, one might argue that neo-Darwinism is, in essence, in intent, an axiomatic theory, with initial hypotheses and all else following more or less (usually less!) rigorously. Or, one might argue that it is not. If one thinks that the theory is in outline axiomatic, then one must explain why the ideal is not yet realized. Usual excuses include the complexity of the task at hand, the theory's relative youth, and so forth (Ruse, 1969b, 1972a,b). If one denies the relevance of the ideal, then one must forward some other model of understanding. Usually it is suggested that the evolutionist, like a historian, is "telling a story" or "narrative," in which tight connections are inappropriate (Gallie, 1955; Goudge, 1961; Ayala, 1970b, 1971; Thompson, 1983b,c).

One advantage to going with the axiomatic-as-ideal option is that it readily takes account of the history of Darwinism (Ruse, 1973a, 1977b; Caplan, 1979; but see Hull, 1974a; Beckner, 1959; Suppe, 1977). Darwin himself certainly aspired to an axiomatic theory. Moreover, such an option makes much sense of what I shall emphasize as the important role of population genetics in evolutionary (neo-Darwinian) thought. However, I do think differences of opinion reflect (in part) a difference in judgement about what the philosophers should be doing. Should they be saying (with the first alternative) what science *should* be like — should they be *prescribing*? Or, should they be saying (perhaps with the second alternative) what science *is* like — should they be *describing*?

Obviously, in a way the philosopher who is worth his or her salt simply has to do both. One has to stay in touch with the way that science really is — otherwise, one's philosophy becomes quite sterile. Yet one has to have some standards, some way of evaluating good science from bad science, or else, one simply ends up as a rather bad journalist.

Nevertheless, different people do have different inclinations.

This point is brought out clearly when one considers the options which present themselves, if one decides that an axiom system is indeed the ideal for evolutionists (neo-Darwinians). Does one pull back and argue that modern theory, even as it stands today, shows the importance of axiomatization? This is what some would argue, without feeling the need to formalize the theory further. As with the Kingdom of Heaven, knowing that there is an ideal is enough for the messy, real world. Or does one push one's prescriptivism all the way, arguing that axiomatization is the only appropriate end for good science, and that therefore one had better provide just that for the evolutionist?

Such an extreme move as suggested in this second disjunct has been made by Mary Williams (1970, 1973) and has been endorsed recently by Alexander Rosenberg (1985a). Williams provides an axiomatization of her own devising, in which, by starting with claims about such things as limited resources and consequent struggle, she can get derived theorems about the effects of selection. This is an impressive achievement, although it has not been without its critics (Ruse, 1973a). In order to axiomatize, Williams has stayed at a high level of generality. For instance, her system applies indifferently to biological entities as broadly apart as genes and species. Depending on one's interests, at one level one can get the Hardy-Weinberg law and at another level something of the order of the Lotka-Volterra equations. One may look upon such generality as a strength. Given some of the trends in philosophy of biology (to be explained shortly), probably more people today than twenty years ago would prize such generality. However, others continue to fear that the generality is purchased at the price of saying anything of specific interest. Is the descent of an ecosystem even remotely like the descent of a gene? A critic might also note that practicing biologists seem not to have embraced Williams' system with much enthusiasm. (For more general discussion on these various issues, especially about the relevance of axiomatization, see Caplan, 1986; Mayr, 1985a; Munson, 1975; Olding, 1978; Riddiford and Penny, 1984; Brandon, 1978b; Sober, 1983a, 1984a, 1987a; Wasserman, 1981; Grene, 1981. For more specifically on Williams, see Williams, 1981; and very critically, Jongeling, 1985.)

I come to the second question posed above. Is modern (neo-Darwinian) evolutionary theory genuinely empirical? Does it tell us about the real world, "out there"? Or does it merely redescribe, in fancy language, that which we always knew? A surprisingly large number of people have problems on this score, and this worry affects not merely philosophers but extends right into the scientific community itself (Peters, 1976; Manser, 1965; Lewontin, 1977; Platnick and Gaffney, 1978; Patterson, 1978a,b). For the moment, I will stay with objections to and defences of neo-Darwinism, leaving alternatives until later. Let me start with the critique of Karl Popper, unquestionably one of the world's leading philosophers of science. He has been a paradigmatic doubter, for he has argued that Darwinism is no real science — that it cannot be properly falsified. It is rather a "metaphysical research programme" (Popper, 1972, 1974). It is true that, for Popper, "metaphysics" does not quite have the pejorative connotations that it has for some other philosophers, notably the logical positivists; but such a label is hardly a mark of respect (Popper, 1959, 1963; Ruse, 1977a).

Much of Popper's unease with Darwinism — an unease shared by many others — has centered on the key mechanism of natural selection, especially under its alternative name of the *survival of the fittest*. Who is it who survives (more strictly, who is it who reproduces)? The answer seems to be that it is those that biologists identify, *by definition*, as the "fittest"! If white moths survive predators, and green, black, and blue moths get eaten, then the white moths are the fittest. But this then means that the key mechanism of the Darwinian reduces to the empty tautology that those which survive are those which survive. This may be true, but it is hardly the foundation of a genuine empirical science (Lewontin, 1969; Ruse, 1971b).

To be honest, for anyone versed in postwar Anglo-Saxon philosophy, this natural-selection-as-tautology critique has a slightly old-fashioned air to it (Sober, 1984a). Thanks to the work of philosophers like W.V. Quine (1953), the belief that claims can be divided, unambiguously, into tautologies — (analytic connections between ideas) — and empirical statements — (describing facts about the physical world) — has been thrown into grave doubt. It is just no longer obvious that one can dismiss something simply by slapping the label *tautology* upon it. But, this point apart — and I shall return to it shortly — defenders of neo-Darwinism have argued that in at least three different ways claims about natural selection transcend the formally empty. (See also Beatty, 1984; Byerly, 1986; Dawkins, 1983; Stebbins, 1977; Waters, 1986a; Caplan, 1977, 1985.)

First, there is the presupposition that a struggle is occurring. If organisms never died and never reproduced, there would be no selection. Similarly, there would be no selection if all organisms reproduced asexually once, at the same time, and then and only then died. This may all be "obvious," but it is true and empirical, as also is the second point, that organisms differ (and going beyond selection to evolution, that such differences are heritable). If every living thing were identical, there would be struggle without selection.

Third, and most crucially, there is the claim that the sorts of features which help in the struggle on one occasion (that is, which contribute to fitness) are also the sorts of features which help on like occasions. The point is that the world runs in a regular kind of way and that a helpful feature (an "adaptation") has a kind of universality to it. If this were not the case, there would be nothing *systematic* about selection, systematic being a feature evolutionists always stress (Wimsatt, 1980b; Brandon, 1978a,b; Waters, 1986a). Take, for instance, the classic example of selection in humans, the maintenance in many populations of genes for malarial resistance, despite the fact that such genes, when possessed in a double dose, homozygously, bring on sickle-cell anaemia. Absolutely central to this explanation is the empirical claim that genes which have these effects in one place or time will have similar effects in other like places or times. Such a claim may be false, but that of course is a risk one takes when one is dealing with nontautologies (Ruse, 1981b).

Of course, this is not to say that what evolved as an adaptation always helps. If circumstances change, then so might the virtues of adaptations. It is just that natural selection depends on the regularity of similar causes leading to similar effects. (This is all very close to a commitment to the value of inductive generalizations, and given Popper's well-known antipathy to induction, might account for some of his troubles with selection. Although, see Popper 1978, 1980, 1984, for revisions and recantations.) Note that, as I have unpacked selection, fitness comes out as a funny kind of property like a disposition or a propensity. A plate glass window has a propensity to break — not all of the time, or even every time a brick is thrown at it, but on a certain proportion of occasions. Similarly, if one speaks of the fitness of, say, the white moth, one is not saying it will always survive, but rather that it has a propensity calculable in terms of survival rates (Brandon, 1978b; Mills and Beatty, 1979; Brandon and Beatty, 1984; Sober, 1984a,b,d, 1987b; Beatty, 1987a,b; but see Rosenberg, 1982a, 1983a,b, 1985a, 1986a; Rosenberg and Williams, 1985, 1986).

Yet still the critics profess their unease with neo-Darwinism. Surely it fails as a genuine theory (that is, as a theory akin to those found in the physical sciences), if only because it deals always with events after the fact. It is a historical theory. Hence, apart from the looseness and so forth, even in principle it gives no scope for predicition (Popper, 1974; Patterson, 1978b). It does not even try to predict the fate of the elephant's trunk, for "future history" is a contradiction in terms. Thus, still there can be no genuine test. It remains unfalsifiable. It is not real science.

Defenders of neo-Darwinism have made several responses to this line of argument. Most importantly, they point out that there does seem some confusion about what the evolutionist is trying to do. It is perhaps useful to make a threefold distinction. First, there is the attempt to establish evolution as *fact*. Do organisms have a history of change and development, perhaps from "one or a few forms"? Second, there is the attempt to establish evolution as *path*. What course did organisms take from past to present? Did birds come directly from ordinary reptiles or via the dinosaurs? What about "phylogenies"? Is evolution progressive (Nitecki, 1988)? Third, there is the attempt to give the *mechanisms* of evolution. Was natural selection, for instance, all-powerful (Ayala, 1985a,b; Ruse, 1984c)?

Now, as far as fact and path are concerned, there certainly is no attempt at prediction. In themselves, they are not the sorts of things which make for predictions. But in comparable parts of physics, there are no attempts at prediction. That the earth was formed 4 1/2 billion years ago is a fact, and that is that. So questions of prediction and falsifiability only really come when we are dealing with mechanisms, although of course we may be using theories (which are falsifiable) when we are trying to establish facts and ferret out paths. However, here, so the defenders continue, we find that neo-Darwinism is not so very much nonpredictive, particularly when we remember that "prediction" in science often involves inferring backwards that something has happened. (This is sometimes known as "retrodiction.") Let me simply say here that, certainly with respect to short-term experiments, a great deal of effort is put into showing how neo-Darwinism forecasts results — with some success, argue its defenders (Alexander, 1979; Dawkins, 1986; Endler, 1986; Futuyma, 1979; Ruse 1982a).

Philosophically, what is at least as interesting is the fact that many would claim that there is more to good or genuine science than prediction and consequent falsifiability (Kitcher, 1983; Ruse, 1984b,c) — although I myself would not downgrade Popperianism to the extent that many of my fellow philosophers would. Most positively, in the best sciences — certainly in the best sciences of physics and chemistry — a major mark of why we think the scientist is saying something significant about reality is because of the way in which unification is effected. Good theories exhibit what William Whewell (1840) called "a consilience of inductions." They bring together disparate pieces of evidence, binding them into a whole, until we exclaim: "Such convergence cannot be mere coincidence. It must be a reflection of reality!"

Picking up then on the third question posed above: Is Darwinian theory consilient? Does it bind together different parts of biology so that together they make a convincing empirical whole? As many commentators have noted, one of the most distinctive things about modern evolutionary studies is the way in which so many different areas do come together (Beckner, 1959; Ruse, 1973a; Dobzhansky et al., 1977). One finds systematics, embryology, paleontology, genetics, biogeography, and much more — all dealt with (more or less adequately) in any competent evolutionary survey. But do we have genuine consilience? There has been dispute about how the various parts mesh together. Some have argued that one merely gets a diffuse net, as it were, with particular subjects hanging on at various intersections (Beckner, 1959). Some have even gone so far as to argue that this looseness shows a crucial weakness in the evolutionist's thought. Individually the various subjects barely stand upright. Taken together, they collapse entirely (Himmelfarb, 1962)

My own feeling is that the critics and doubters do evolutionists an injustice, although perhaps not so much an injustice as neo-Darwinian evolutionists sometimes claim (Ruse, 1973a; Schaffner, 1980). One does have a perfect consilience, with various subdisciplines (like paleontology) all converging on a core. Together the parts may be weak, but mutually they are reinforcing. Why is the guilt for the murder pinned on the butler? Taken individually, the clues are not convincing. Taken together, his guilt is "beyond reasonable doubt." Likewise, the Darwinian evolutionary system is made genuine empirical science by the findings and theories of its parts. Paleontology, biogeography, and systematics furnish the bloodstains, the fingerprints, the broken alibis. Put together, the system is "beyond reasonable doubt." (However, for more discussion on these various points, see also Caplan, 1978b; Lloyd, 1987a, 1988b; Michod, 1981; Milkman, 1982; Pollard, 1984; Selander, 1982; Darden, 1986a; Templeton, 1982; Wassermann, 1978; Wimsatt, 1987.)

Yet, having said this, there is often confusion in Darwinians' minds about whether they have established evolution as fact or evolution as mechanism. This is a confusion which goes back to the *Origin*, for Darwin himself twisted together his arguments for the fact of evolution and for his mechanism of natural selection. Certainly, the consilience establishes the fact — as much as anything can be established in science. The mechanism perhaps requires more argument. Which requirement starts to push us toward the separate parts of the neo-Darwinian synthesis. Surely, above all else we must look at the core of the Darwinian argument — the center of the consilience, which is supported by and in turn informs the other disciplines within the overall theory (Darden, 1986a). But what is this core? Obviously, it has to be that part of biological science which focuses on natural selection and on its effects on groups through time and under various circumstances. This is the subject known loosely as "population genetics" (Dobzhansky, 1970). Clearly, therefore, it is to this subject that we should turn next in our philosophical inquiry. As we shall see, there have been differences of opinion, as striking about the core as about the whole theory. (See Rosenberg, 1980a, 1985a, for very different thoughts on the structure of neo-Darwinism.)

CHAPTER 2

POPULATION GENETICS

B egin at the beginning. As noted, Darwin had no true grasp of the principles of heredity (Mayr, 1982). Consequently, he thought of selection exclusively in terms of organisms. One organism does better (or worse) than another. Since the *Origin*, the evolutionist's greatest advances have been in an ever-deeper understanding of the rules behind heredity. In particular, the gene-as-unit-of-function — that which causes physical (phenotypic) features — has been brought to the fore. Does this mean that selection should now be considered exclusively in genetic terms — as change in gene ratios? Some scientists and philosophers have certainly implied so, and some have indeed said so (Ayala and Valentine, 1979). Others deplore what they feel is an undue disregard for the actual living organism, which is, as they point out, that which actually has to survive and reproduce (Mayr, 1963).

We can perhaps go some way toward resolving this issue if we make use of a distinction proposed by Richard Dawkins (1978, 1982). He would have us distinguish between "replicators" (notably genes), which persist through time and which (or copies of which) get transmitted, and "vehicles" (notably organisms), which contain the replicators and which actually battle in the struggle. At one level, selection is between vehicles. But the causal effect — a feedback effect because the replicators make the vehicles what they are — is selection between replicators (Dawkins, 1976; but see Gould, 1980a).

Certainly it is the case that, conceptually, evolutionists in this century have tended to think in terms of replicators, of genes, believing that once their dynamics are worked through, the rest of the story at the physical level can then be fleshed out (to use an appropriate metaphor). What this means, therefore, since selection is a mechanism which can only work *between* organisms — that is, when there are groups — is that evolutionists must first derive claims that apply to genes in groups or populations, and then introduce factors of change like selection. This, of course, they do, through deriving the Hardy-Weinberg law (a claim about genes in populations) from the basic Mendelian laws of transmission (claims about how genes get passed on from one individual to another).

At once, philosophical questions start to intrude. Most particularly, if we grant that population genetics is at the core of evolutionary theory, can we now go further toward answering the major question posed in the last section, namely, whether or not the axiomatic ideal is appropriate for evolutionary theorizing? That one does find formal inferences, as from Mendelian laws to the Hardy-Weinberg law, cannot be denied (Ruse, 1973a). But is this enough to make sweeping claims? Some feel that, given the nature and importance of population genetics as it exists now, this is enough to speak of evolutionists having the same formal ideals as physicists. Others argue that, interesting and significant though population genetics undoubtedly is, it is really quite too unformed and diffuse, especially when considered with the rest of evolutionary studies, to be classed with such a science as, say, Newtonian mechanics (Rosenberg, 1980a, 1988b).

Again, therefore, we find the prescriptive-descriptive dichotomy raising its head. There are those (prescriptivists) who have argued that much greater rigor needs to be achieved before one can talk confidently about the appropriateness of axiomatization (Woodger, 1937, 1952; Kyburg, 1968). More interesting, perhaps, are the descriptivists, especially those who acknowledge the importance of population genetical thinking but who still feel that the axiomatic ideal is misleading. Essentially, their worries center around the great difficulty that one has making anything remotely resembling a general statement in biology (Beatty, 1978, 1981). As soon as one makes a claim, say, that white moths, on average, outreproduce black moths, then exceptions start to flood in. But what does this problem imply? Either one ignores the exceptions, or not. If one ignores the exceptions, then, to be honest, one's science is hardly true. More particularly, one's axiom system is no longer composed of claims of the kind one finds in the physical sciences — true, universal statements about the way the world must work, namely, "laws." If one acknowledges the exceptions, then axiomatization becomes impossibly difficult, and in any case one has merely claims of limited scope. These may well be true, but they are hardly lawlike (Smart, 1963; Thompson, 1983c).

Some of the worries of those who incline this way may well be misplaced. For instance, doubts are expressed about the Hardy-Weinberg law. Can this really be a genuine law of nature? After all, it is only the binomial distribution, which follows because the law is a straight generalization of the Mendelian laws. Nevertheless, to cavil at the law or its qualifications is to miss the nature and significance of the law (Ruse, 1973a; Sober, 1984a). On the one hand, Mendel's laws are unambiguously empirical. On the other hand, the Hardy-Weinberg law sets background conditions for what would be the case if there were no disruptive forces (like selection or sampling effects due to finite numbers). It is an equilibrium law exactly analogous to Newton's first law of motion. Both laws give us a presumption of stability, against which the forces of nature can work out their drama.

A more significant worry of the descriptivists points to the fact that evolutionists themselves seem unconcerned with building a grand axiom system that applies (as one person has put it) everywhere and everywhen (Smart, 1963). Rather, one finds efforts directed toward the construction of restricted *models* whereby certain initial boundary conditions are specified and then the specific consequences are traced (Lewontin, 1974; Ruse, 1977b; Giere, 1979). Some biologists put their effort into model building, some into studying nature, and some into bringing certain models to bear on certain empirical situations. There is, therefore, an ongoing dialogue between the theoretical and the empirical. But, there is no attempt at making one general system.

Descriptivists who find this point definitive, therefore, argue that model rather than general system building is the key process for the evolutionist. Such descriptivists claim that they can thus acknowledge the use of formal methods in biology, but they can also stay true to what biologists actually do (Beatty, 1978, 1980, 1987c; Brandon 1981a; Griesemer, 1984; Thompson, 1983a, 1986, 1987, 1988a,d; Lloyd, 1986a,c, 1987b, 1988b; Richardson, 1986a,c; but see Sloep and van der Steen, 1987a,b). Moreover, these philosophers usually make their claim even stronger by arguing that, far from separating biology from physics, they bring the two closer together in spirit. They argue that in physics also, the building of restricted models, rather than the construction of sweeping axiom systems, is the norm. In arguing thus, these descriptivists rely on the conclusions of a like-minded group of philosophers of physics, the so-called semantic school (Suppe, 1977; van Fraassen, 1980).

The debate still continues between supporters of axiom systems, often labeled partisans of the "received view," and the semantic theorists. One strong point favoring the semanticists is the way in which worries about the empirical nature of natural selection are readily quelled. First, in line with Quine's point about the lack of a sharp distinction between analytic and nonanalytic, the semantic interpretation supporter notes that *any* model, including populational genetical models using natural selection, is definitionally true until it is applied to reality, when it becomes empirically true or false. Hence, at one level it is meaningless to ask if selection is a tautology — it all depends on the context.

Second, it is noted that some applications of some models using selection definitely point to real empirical processes — the sickle-cell anaemia case, for instance. What the semanticist seizes on is the fact that all evolutionists, when defending selection, work on a case-by-case basis — which is just what the semantic philosophy predicts.

Going the other way, and noting a major worry with the semantic position, I would point out that if one downplays the importance of laws in scientific activity, then one is at least blurring (if not fudging) the distinction between science and nonscience. Perhaps there is indeed no hard line of demarcation. But if there is no line at all, then the way is open for critics of biology to argue that their own rival views have equal claim on our attention — more particularly, that their views have equal claim in our attention in class. Significantly, one person (a practicing biologist, in *Nature* no less) has already argued that the semantic view opens the way to a justified belief in miracles (Berry, 1986).

There are many other philosophical avenues opened by the study of population genetics. Some will be raised shortly (Sober, 1984e, collects many of the seminal articles). For once, let a note of pride (as a philosopher) creep into my voice when I say that I think some real contributions have been made, both in locating the importance of population genetics and in explicating its nature for evolutionary studies. This success is indicated by the fact that we are now finding biologists and philosophers jointly authoring papers on the subject. (For instance, Sober and Lewontin [1982]. Incidentally, given its generality, Williams' [1970] axiom system might mesh nicely with the semantic position.)

CHAPTER 3

MOLECULAR BIOLOGY

In 1953, James Watson and Francis Crick published their paper on the DNA molecule, and the world of biology has never again been the same. In one dramatic stroke, the physicochemical sciences were brought right into the domain of organisms, making clear and understandable much that hitherto had not been grasped at all. With progress came resentment, for conventional biologists felt insecure and threatened when their lives' work was dismissed as irrelevant (Watson, 1968). Thirty years later, it is hard not to laugh at what has happened. Molecular biology thrives. That was to be expected. But so also does conventional biology — as never before. Moreover, conventional biologists appear very comfortable with molecular biology, sending their students to take courses in it and calmly beginning their books (particularly their textbooks on evolutionary theory) with detailed expositions of the DNA model and its implications.

There is something interesting going on here, something interesting about the way in which science develops and expands. Naturally enough, therefore, the molecular-conventional biological interface has attracted much philosophical attention. In order to unravel the discussion, let me focus on the term *reductionism*, since it was this that traditional biologists claimed to find so unwelcome about the growth of molecular biology (Mayr, 1969b). If we can see what people might or might not be claiming, we may learn what might or might not be at issue in the meeting of the molecular and the organic sciences (Wimsatt, 1979). To go some way toward avoiding ambiguity, I shall use a threefold division proposed by Francisco J. Ayala (1974b). That Ayala is one of today's leading evolutionists, at home on both sides of the molecular-conventional interface, gives a certain authority to my choice (Ruse, 1988c; Rosenberg, 1985a, 1988a).

First, there is the thesis of *ontological* reductionism. This is the claim that bigger entities are composed of no more than smaller entities. The whole is composed of nothing but its parts. One does not have the parts plus some mysterious new substance. A tent is nothing but canvas and the ropes and the brass fittings and the pegs. An organism is nothing but the molecules of which it is made. There is not some lifeforce or suchlike thing which is needed to make a living being. In this sense — and, as far as ontological reductionism is concerned, in this sense only — humans are like machines.

With the decline of pure vitalism, I doubt there is anyone who would want to deny this kind of reductionism, at least not in any way that need trouble us here. Certainly, at the interface we are now considering there seems little temptation to suggest, say, that the Mendelian gene has a life force denied by the DNA molecule. When we come to humans and to their relationship with their biology, perhaps we shall find questions of ontological reductionism a little more problematic. (But see Caplan, 1987, 1988.)

Next we have *methodological* reductionism. Here the claim is that, all other things being equal, one ought to explain in terms of the smaller rather than the larger (Williams, 1966, 1985a). Explain sex in terms of the chromosomes, if you must, but, even better, explain in terms of the DNA molecule. Methodological reductionism goes beyond ontological reductionism, not only in the claim that the bigger is made up of the smaller but also in the claim that reference to (and only to) the smaller gives greater scientific understanding than reference to the larger. Presumably "understanding" here refers to explanation, predictive fertility, unification, and so forth.

Methodological reductionism is significantly more philosophically contentious than ontological reductionism. I take it that no one would deny that a major trend — probably *the* major trend — in science has been toward the explanation of the large in terms of the small. Moreover, even before the coming of molecular biology, methodological reductionism of this kind was being practiced highly successfully in biology. Think of the importance of the classical concept of the gene. This is not to deny — what is often pointed out — that there is usually much more to explanation than merely the positing of tiny entities (Mayr, 1969b, 1975, 1985; Campbell, 1974b). One has to take into account the way the entities function and work, and the kinds of ways they are distributed and organized. This is particularly important in biology. For instance, the positions of genes on chromosomes can make all the difference to phenotypes (Hull, 1974a).

On the other hand, it should be pointed out that organization is important in physics also (Ruse, 1973a). It makes all the difference to the effects whether one has one's resistors in parallel or in series. Furthermore, paradoxically, given that so many of the comments about the significance of organization are made by those suffering from an extreme case of physics fear, no one is more sensitive to *biological* organization than the molecular biologist. After all, he or she argues that his or her main focus of interest, the DNA molecule, is made up of four basic building blocks. Differences are all a matter of the way one rings the changes. Shuffle the units one way and one gets a mouse. Shuffle the units another way and one gets an elephant (Grene, 1987).

But is methodological reductionism in biology an overridingly good strategy, to be pursued just as far as possible and to be abandoned only reluctantly when one reaches a dead end? Or is a total commitment to methodological reductionism a bad policy, philosophically cramping and leading to thin, arid science? We shall shortly meet some very strong opponents to methodological reductionism. At the molecular-conventional interface, my sense is that philosophers (like biologists) generally think it should be encouraged; this statement applies even to those who generally react strongly against methodological reductionism. I suspect that there is an obvious reason that such reductionism causes no controversy at this interface, namely, that already with Mendelian genetics one has committed oneself to methodological reductionism. One is trying to explain physical properties in terms of underlying microentities. To push all the way to the DNA molecule is only to take to the limit what one has already accepted as a good way of doing science. The objections rise when one argues, in the first place, that reduction and the genes, of whatever order, offer the best path for the biologist. Molecular biology simply intensifies these worries. (See also Doolittle, 1985; Hunkapiller et al., 1982; Jungck, 1983; Robinson, 1986; more generally, see Wimsatt, 1974, 1976a,b, 1979.)

Putting aside these issues for one moment longer, let us turn now to our third kind of reductionism: *theory* reductionism. Although it is probably methodological reductionism which has most engaged scientists recently, it is theory reductionism which has spilt most philosophical ink — and, from a broader perspective, it brings us back to the ultimate aims of science, most particularly of biology. This kind of reductionism centers on the connections between theories, usually between an older theory and a newer rival (Darden and Maull, 1977; Maull, 1977). Given, as in the case of the older conventional biology and the newer molecular biology, that there are often tensions between the older and newer, what in fact is the formal relationship between the two theories?

Two quite different answers have been provided by philosophers. The reductionists see continuity (Nagel, 1961). The older theory is not so much displaced by the newer theory as absorbed within it. The alternative philosophical view denies that there is this continuity. The newer theory simply pushes the older theory out of the way. It is finished. It is over — except, perhaps, as an aid to classroom teaching. It is replaced (Kuhn, 1962). In the light of these alternatives, how should we interpret the relationship between the older conventional biology — in this context, we are specifying genetics — and the newer molecular biology — again, specifying genetics? Some argue that we have as neat a case of reduction as we could hope to have (Schaffner, 1967a,b, 1969a,b,c, 1974, 1976; Ruse, 1973a, 1976a). Others argue that there is nothing remotely like reduction, and harmony has been achieved only because conventional biology has bent with the wind, abandoning its own ideas whenever the molecular invaders approach (Hull, 1972, 1973b, 1974a, 1976b, 1979a; Wimsatt, 1976b; Grene, 1974; van Balen, 1987).

The problem arises over the exact interpretation one gives to the reductionist's notion of "absorption." Originally, as the approach was explicated by Ernest Nagel, the idea was that the older theory would be shown to be a deductive consequence of — and perhaps a special instance of — the newer theory. In order to effect this argument, particularly since the language of the two theories was often different, one would need certain connecting rules identifying items with one theory with items in the other. Unfortunately, however, in the case of genetics — and it turns out that this is not a unique instance — it seems that one cannot get appropriate connecting rules (or "bridge principles"). One trouble is that, whereas the Mendelian gene is an all-purpose entity which can be passed on, which mutates, and which carries the information required for growth, the molecular geneticist tends to divide up these various activities, ascribing them to different amounts (and, sometimes, parts) of the DNA molecule. At best, therefore, it seems that one might deduce from molecular genetics a "corrected" version of traditional biological genetics.

Debate centers around how much correction one allows before one drops talk of reduction in favor of replacement. Let me simply content myself here with two general comments. First, it is clear that if there is anything which even remotely approaches a reduction — and at best this approach is toward an ideal of the most abstract kind, for no ambitious molecular biologist ever wants to pause to fill in the steps — then it is a great deal more complex than was implied by philosophers, even a mere ten years ago (Rosenberg, 1978, 1985a; Kitcher, 1984a; Kincaid, 1987). The links and cross-links required to show how nonmolecular genetics can be connected to molecular genetics must be of an order of magnitude and complexity rivaling the lines of a major telephone exchange. There is so much richness at the molecular level, and at the nonmolecular level, that showing the latter a logical consequence of the former would be an undertaking of horrendous theoretical and practical difficulty.

But, second, the reductionist's ideal is not simply a philosophical fantasy, sought only by those with no real feeling for biological science. There is motive behind the drive, even if the method seems a little shaky at times. If, in fact, older theories are in some sense special cases of newer theories, then (as seems now to have happened in biology) one can see why those who have worked with older theories — more specifically, those who have wrestled with the problems thrown up by older theories — might welcome newer theories. The later arrivals offer new ideas and techniques, more powerful than before, which can be used on older, thus-far incalcitrant problems. The worth of an older theory's concerns are not denied. It is just that new avenues of attack are opened up.

Certainly, if there is any truth in this general scenario, then some of the moves made in biology — thinking still specifically about genetics — make sense. At the general level, one can see why evolutionists now start discussion with the DNA model. Their previous work, with Mendelian genes and the Hardy-Weinberg law and so forth, is not thrown overboard, but is rather given a deeper backing through our new knowledge of the molecular basis of life. And more specifically, it becomes clear why certain projects are attempted and (if completed) hailed as significant.

Illustrating this latter point, I will refer to a major dispute that separated premolecular biologists, right from the time when the modern theory was put together in the 1930s (Ruse, 1976a; Beatty, 1987a,d). Do populations of organisms, as they occur in nature, contain a significant amount of variation, or not?

Theodosius Dobzhansky (1951) claimed that they do — and moreover, that natural selection (through such effects as balanced heterozygote fitness) is responsible for keeping such variation high. H.J. Muller (1949) claimed that they do not — and moreover, that natural selection (through a kind of cleansing action) is responsible for keeping such variation low.

This was no minor dispute. On its resolution rested much of the plausibility of modern evolutionary theory — a plausibility quite missed by many critics, like Popper (Ruse, 1982a). If there is always plenty of variation held in populations, then when needs arise, selection can at once swing into action. There is no need to wait for new appropriate variation — a need which many critics feel that neo-Darwinism fails to supply, given that Darwinians always emphasize that new variations are "random," in the sense of not occurring according to need.

One could not solve the variation problem by traditional means. Factors which could make major differences in the evolutionary scale of things were too slight for conventional measurement. Enter molecular biology. People like Richard Lewontin (1974) were able to express genetic differences in molecular terms and to show that there is indeed massive variation always held in virtually every natural population. Dobzhansky was vindicated and neo-Darwinian evolutionary theory made that much more plausible. (See also Ayala, 1972b, 1974d, 1975a; Ayala et al., 1974.)

As so often happens in science, there is rather more to the story than this, but not such as to affect the point just made. If a theoretical reduction occurred — admittedly something which would be most difficult to spell out in full — then the sorts of effects that molecular biology has had make sense. Otherwise, one has to find some other analysis for the ways in which molecular and conventional biologists seem to have found a meeting point.

It seems, therefore, that there is something to be said for theory reduction, although its exact specification is apparently a task of great complexity. And with such an ambiguous conclusion, the time has come to call complete this brief survey of the problems of reductionism as they arise in the coming of molecular biology to the problems of genetics. In a way, more questions are asked than solutions offered. But at least we are getting a better grasp of some of the challenges, and that is progress. (For more recent, sophisticated discussions of reduction, see Balzer and Dawe, 1986a,b; Kitcher, 1982; Kimbrough, 1979; Rosenberg, 1978, 1985a, 1988a; Caplan, 1981c, 1987, 1988; Tennant, 1985, 1986, 1987; Brandon, 1985b,c,d; Dupré, 1983. For historical and detailed scientific perspectives, see Burian, 1985; Bechtel, 1982a,b, 1983, 1986a,b,c; Darden, 1980, 1986a,b. For semantic theorists' opinions, see Beatty, 1983; Thompson, 1988f.)

CHAPTER 4

CHALLENGES
TO THE
PARADIGM

I want now to change course, somewhat. Thus far, I have been working with what, without prejudice, I will call the "standard picture" of evolutionary theory, even as we have expanded out to molecular biology. But Darwinism — and neo-Darwinism — have always had rivals, and although some twenty or so years ago the dissenters were few (and often beyond the edge of accepted science), in recent years the voices of criticism have been rising. Moreover, for reasons about which I shall speculate shortly, we find that philosophers of biology have been much involved in questioning the orthodox paradigm. In this section, I shall run the gamut through increasing degrees of opposition to neo-Darwinism, concentrating as always on the philosophical issues and involvement. (Brooks, 1983 is a good survey of the biological ideas.)

To start right at the beginning and to set the scene, let us take one who is not at all critical of neo-Darwinism. To the contrary, the English biologist Richard Dawkins (1983) has recently taken to arguing that if one adopts an evolutionary stance, then one *must* accept natural selection as the main mechanism. Moreover, Dawkins insists that selection must be seen as a process centering on the individual rather than on the group. Scientific alternatives, like Lamarckism (inheritance of acquired characters) or saltationism (evolution by jumps) or orthogenesis (trends fueled by unseen forces) or group selection (that is, selection producing adaptation for the group and against the individual) he dismisses as simply inadequate, in principle. (But see Cullis, 1984; Steele et al., 1984.) It was Dawkins who coined and popularized the notorious metaphor the *selfish gene* although (using terminology already introduced) he would now have us distinguish between replicators, where these would be individual genes passed on from generation to generation, and vehicles, where these might be organisms which battle for supremacy in life's struggles.

To be candid, even supporters of neo-Darwinism might be wary of help like this. A priori demonstrations of the necessity of natural selection hint that perhaps the tautology objection could be let in by the back door. However, if one looks at Dawkins' arguments, one finds that his position is not as starkly divorced from the real world as this, and he does make some interesting assumptions. (See especially Dawkins, 1986.) In particular, on the one hand, regarding the necessity of natural selection per se, we find that Dawkins relies openly on the empirical assumption that the most dominant feature of the organic world is its ubiquitous adaptedness. Jocularly, he refers to himself as a "transformed Paleyist," meaning that like the natural theologian he thinks the mark of the living is adaptive functioning. On the other hand, with respect to the focus of selection, we find that Dawkins makes heavy commitments to what he takes to be simplicity or parsimony. To locate selection at the lowest possible level is, for him, a principle of good science. If faced with a choice of the bigger or the smaller, the smaller is always preferred.

Both of these assumptions have been queried. Taking the later first, as is well known the whole question of the level(s) at which natural selection works has been the subject of intense biological scrutiny in recent years. No one wants to deny that the emphasis on individualistic approaches to selection has paid great dividends. One thinks, for instance, of William Hamilton's (1964a,b) brilliant work on the hymenoptera. Nevertheless, an approach is one thing, near-categorical instance, is another. Following the strong individualistic stand taken, first by such people as George C. Williams (1966) and John Maynard Smith (1983a, 1984a), and then by Dawkins (1976, 1982), theoretical models have been devised showing circumstances in which group effects might obtain (Wilson, 1980, 1983). In addition empirical studies, particularly by Michael Wade (1978) of Chicago, have tried to show how the group might be the center of selection. (Actually, Darwin was the first evolutionist to take a strong individualist stand: Ruse, 1980. Griesemer and Wade, 1988, contains an excellent discussion of the problems with testing for group selection.)

Philosophers too have been looking at the levels-of-selection question (Brandon and Burian, 1984). In the first instance, they have been questioning assumptions about simplicity, and whether these are necessary for science — or, more importantly, necessary for the best science. Perhaps such assumptions cut one off from the other desiderata, like predictive fertility. In the second instance, there has been work on the exact meaning of "individual" and "group" selection — work the effectiveness of which is acknowledged by Dawkins' recognition that he needed to distinguish between replicator and vehicle. There is growing realization that in the inquiry about the nature of selection one must at least separate out cause and effect. Selection might always take place as a causal mechanism between individuals of one kind (say, organisms, or perhaps exceptionally genes). The effects of such selection might, however, rest in entitles at different levels, say, between groups (Sober, 1984a).

Obviously (and explicitly) what we have is another dispute

over the worth of methodological reductionism. Someone like Dawkins believes that it is a good approach which has led to impressive results. Had evolutionists not tried to explain using the lowest levels of selection, then (for instance) none of the powerful tools of game theory (as represented through work on "evolutionarily stable strategies") would have been opened up for them. Others argue that such reductionism is impossible in principle and misleading in practice. For instance, biologist Richard Lewontin and philosopher Elliott Sober show how much populational thinking, even using conventional biological genes, cannot explain simply in terms of the smallest units (Sober and Lewontin, 1982; Sober, 1981c, 1984a). The case of sickle-cell anaemia, a by-product of balanced heterozygote fitness, proves the point. To ask if the sickle-cell allele is fitter than the wild-type allele is meaningless, for only when considering the genes in combination, as heterozygotes or homozygotes, do the questions make sense.

On the basis of such examples as these, Lewontin and Sober conclude that the drive to methodological reduction should be viewed with suspicion. The reductionist is almost forced into thinking that there is something wrong or inadequate with the explanation of the persistence of sickle-cell anaemia, when it is obvious that the very opposite is the case. (For more philosophical discussion of the levels-of-selection issue, see Brandon, 1983, 1985a; Gifford, 1986; Gould, 1980b; Hull, 1980; Lloyd, 1986b, 1988a; Maynard Smith, 1983b, 1984a; Mayr, 1984; Michod, 1984; Mitchell, 1987; Reed, 1978; Richardson, 1983; Sober and Lewontin, 1983; Heyes and Plotkin, 1988; Williams, 1984, 1985b; Wimsatt, 1981a, 1980a; Wilson, D.S., 1984; Darden and Cain, 1988.)

This is an unresolved dispute which we shall encounter again in a moment. A side issue which has emerged comes from the very language used in biology — when used by Dawkins in particular. I have noted that, as part of his case, he makes use of metaphor, most notably speaking of the "selfish gene." Critics have objected that genes cannot be selfish — only people (and perhaps some higher animals) can — and that we have here an unduly anthropomorphic use of language (Midgley, 1985). Defenders of Dawkins concede that perhaps he and others do at times get carried away by his metaphor, but they staunchly defend his right to use it (Mackie, 1978; Ruse, 1979b, 1988e, 1989). They point out that all scientists use metaphors (Hesse, 1984) — *work, force, attraction* — and that such terms have a long and honorable history in biology. What, after all, is the term *natural selection* but metaphor? Defenders of metaphor argue that, like it or hate it, we can never think of selection in the same way again since Dawkins attached *selfish* to *genes*. I will make no final comment. Feelings run deeply on this issue. But what we can surely say is that Dawkins' language does draw attention to the way in which the packaging of science is often as important as the content (Kuhn, 1979).

Turning now to the question of adaptation, this likewise has recently been the subject of intense scrutiny by biologists and philosophers. Most notably, this scrutiny has come through the challenge launched by the paleontologists (especially Niles Eldredge and Stephen Jay Gould), with their theory of punctuated equilibria, which argues that the course of evolution is not gradual and smooth, but goes in fits and starts, with stability ("stasis") broken by swift moves to other forms. Originally, the theory was offered as an outgrowth of neo-Darwinism, using Ernst Mayr's (1963) notion of the founder principle as an explanation of the origin of new species and as the reason that change tends to get forced into short periods. (See Eldredge, 1971, 1982, 1985a,b,c; Eldredge and Gould, 1972; Gould, 1980b,c, 1982a,b,c, 1983a,b; Gould and Eldredge, 1977, 1986.)

Now, however, after toying with thoughts of macromutation, the theory has evolved into a hierarchical synthesis (Arnold and Fristrup, 1984; Grene, 1983, 1985, 1987; Rosen, 1984; Ruse, 1988e, 1989; Salthe, 1985). At one level, that of the individual organism, we get the usual workings of natural selection. Below this, at the molecular level, perhaps we also get selection, although (in line with suggestions of the Japanese evolutionists) drift might be very important here (Kimura, 1983). More significantly, at levels above the individual organism, when we get to groups like species, we find that they too have dynamics of their own. Group effects are important, leading to things like trends, which get exhibited in the fossil record. Although there are sometimes hints of group selectionist thinking, more significant is "species selection," by which whole units work out their destinies (Stanley, 1979; Gould, 1982a). (An example of species selection might come through pressure for larger-sized organisms in an evolving macrogroup, even though the members of individual new species might be smaller and against the trend.)

One thing which particularly marks this theory of punctuated equilibria is the way in which it downplays individual adaptation. The phenomenon is certainly not denied, but it is felt that the Darwinians make too much of it — like Dr. Pangloss (in Voltaire's *Candide*), Darwinians see purpose even when there is none. Against this idea, punctuated equilibria supporters argue that organisms must satisfy "constraints," on which adaptations are hung. Major change is dictated by these constraints, as organisms are switched from one form (or *Bauplan*) to another. The importance of conventional selection is therefore deemphasized. (See Burian, 1983; Gould, 1971; Hoffman, 1983; Kauffman, 1983, 1985; Gould and Lewontin, 1979; Peters, 1983; Reif, 1983; Riedl, 1983.)

A number of philosophers have been attracted to this kind of theory — to her credit, Marjorie Grene (1959) was arguing for something like this some fifteen years before the scientists — and they have joined with Gould (especially) in formulating critiques of adaptationism. Apart from the charge of Panglossianism, a favorite criticism is that Darwinians build "just so" stories. As in Kipling's tale of the elephant's nose being lengthened by pulling, so Darwinians think up all sorts of absurd stories to give adaptive status to every single thing (Gould, 1978; Kitcher, 1985a; Burian, 1981).

As one might by now expect, there have also been vigorous responses to this kind of theory, by Darwinians, biological and philosophical. They argue that problems and excesses notwithstanding, adaptationism is a good strategy, for many things previously thought nonadaptive have since been shown to be under the tight control of selection (Dennett, 1983; Cain, 1979; Caplan, 1981b; Ruse, 1981c,d). In any case, the critics misunderstand the Darwinian case. No one denies that there are some constraints, nor does anyone claim that all features now have adaptive value.

Take the four-limbness of vertebrates, a favorite example of a supposedly nonadaptive *Bauplan*. Perhaps fourness has no value now, but when vertebrates evolved — as fish — having two limbs fore and two aft had strong virtues. Moreover, continuing the critique, Darwinians argue that the notion of species selection is confused (or reduces to individual selection), and the very concept of a hierarchy is jumbled and contradictory. All in all, what is new is not good, and what is good is not new (Maynard Smith, 1981, 1983b, 1987; Dawkins, 1982, 1986; Beatty, 1985b; Thompson, 1988e; Ruse, 1982a; Turner, 1983; Kellogg, 1988; for Darwin's own views see Rhodes, 1986.) See Kitcher, 1987b, for a sophisticated discussion of adaptationism.

Obviously, and again explicit, we have a debate about reductionism, particularly methodological reductionism. Eldredge and Gould and their philosophical supporters believe that it is bad science to try to explain the large, specifically macroevolution, in terms of the small, specifically microevolution. The attempt to do so simply misses important facets of the evolutionary process (that is, evolution as mechanism). Darwinians counter by arguing that the denial of methodological reductionism turns one to pseudomechanisms like species selection. This is not to say that all (or any) Darwinians are blind reductionists in every respect. In a subtle analysis, Francisco J. Ayala argues for methodological reductionism (of macro- to microevolution) but pulls back from full theoretical reduction. Certainly, he allows that there can be no question of inferring the particular paths, or even general patterns, of evolution simply from a knowledge of processes. More knowledge of the broad details is needed (Ayala, 1983a,b,c, 1985a,b, 1987a; Stebbins and Ayala, 1981, 1985).

Perhaps, when the dust is settled from the reductionist wars, we shall find that both sides have moved toward the middle, as Ayala certainly has done. But why is it that so many feel as violently as they do, in opposition to methodological reductionism? And why do philosophers feel this way, suspecting or opposing conventional Darwinism, whether or not they embrace punctuated equilibria theory? Is it simply that Darwinism is inadequate, or that as non scientists — to echo a complaint of Darwin himself — the philosophers fail to recognize how necessary it is to have a theory of some sort, whatever the difficulties? Perhaps both of these points have truth, as also does the point that philosophical opposition (or sense of doubt) is an artifact of the situation. Philosophical inquiry begins when, and to a certain extent only when, scientists fall out. Just as the paleontologists Gould and Eldredge certainly like their theory, in part, because it makes paleontology vital in the understanding of mechanisms, so philosophers like theories which raise philosophical issues.

But there is more. Certainly some of the drive against methodological reductionism by some biologists is linked — quite explicitly — to their Marxist philosophy, which is the epitome of holism. One suspects that this makes the opposition attractive to some philosophers also, particularly given that (as we shall see) the opposition is linked to rejection of Darwinian excursions into the human realm (Levins and Lewontin, 1985; Lewontin and Levins, 1976; Lewontin et al., 1984; Gould, 1979; see also Segerstrale, 1986; Ruse, 1982a).

Nevertheless, with Gould himself I suspect that even the Marxism is less an end in itself and more part and parcel of a general European (as opposed to British) outlook on the world, especially the organic world (Gould, 1983b; Ruse, 1988e). European biologists, back to transcendentalists like Goethe, have always downplayed the importance of adaptation, seeing it as a distinctively British form of natural theology ("transformed Paleyism"!). Europeans have stressed, to the contrary, structure, form, *Bauplane*, hierarchies, and holism (Reif, 1983; Riedl, 1983). What we have going today is a replay of divisions between nineteenth-century biologists (brilliantly documented in 1916 by E.S. Russell in his *Form and Function*). Whereas someone like Dawkins is the direct intellectual descendant of Darwin and (ultimately) Paley, his opponents draw on the continental tradition. As it happens, many North American philosophers of biology, like the biologists, have stronger roots in Europe than in Britain.

If there is truth in this suggestion, then perhaps some light can be thrown on the question which has been raised since the birth of punctuated equilibria theory, (for instance by Stidd, 1980). Does it constitute a new paradigm, a rival to Darwinism, in the sense that Thomas Kuhn introduces the term in his *Structure of*

Scientific Revolutions? The answer is perhaps both "no" and "yes." In one sense, punctuated equilibria theory is not a new paradigm, for it has roots stretching back nearly two hundred years. However, in another sense it is a paradigm which is different from Darwinism or neo-Darwinism. Kuhn stresses that one crucial thing about paradigms is that they make one see things in different ways — and certainly the Darwinians and their critics seem to look in different ways when it comes to adaptation. It is not so much a question of right and wrong, as of different views on things. If there is any truth in this, then possibly — for all of Ayala's valiant attempts at compromise and synthesis — hopes of resolution (between Darwinism and punctuated equilibria theory, and between methodological reductionism and holism) are ultimately doomed (Ruse, 1988e).

Of course, we should not exaggerate the divides between conventional evolutionists and the paleontological critics. Gould (1982a) himself speaks of "expanding" Darwinism, and his philosophical supporters would agree. However, continuing down the road of opposition, we find that in recent years a number of biologists have made more extreme breaks — arguing that the real (or fundamental) processes of evolution have little to do with selection processes of any kind. They turn rather to thermodynamics, and particularly to the non-equilibrium theories of Ilya Prigogine and his followers. They argue that within these theories we find reason to believe that order and entropy can advance together, and this idea, applied to the biological world, can lead to the evolution of species. (See Brooks and O'Grady, 1986; Brooks and Wiley, 1984, 1986; Wiley and Brooks, 1982; Wicken, 1986; Weber et al., 1988.)

These ideas are new and radical. Moreover, there is no universal agreement among the various theorists attracted to the approach. Some want the causal processes confined almost exclusively within the evolving group. Others allow a greater input from the outside. Matters are made more complex by various commitments to certain views on the proper nature of biological classification (to be discussed shortly). Therefore, perhaps it is too early to make definitive pronouncements on these sorts of approaches to evolutionary processes — not that this consideration has stopped conventional students of thermodynamical processes from being highly, almost violently, critical (Morowitz, 1986; and, in response, Wiley and Brooks, 1987).

Nevertheless, such approaches have attracted philosophical attention and praise (Campbell, 1985; Collier, 1985; Depew and Weber, 1985a,b; Dyke, 1985, 1988). Darwin consciously modeled his theory on Newtonian mechanics. It is felt that perhaps it is time to turn to an evolutionary theory which draws on more up-to-date physical science. Some support has also undoubtedly come because of what is perceived as the failure of Darwinism; this support is often coupled with extreme enthusiasm for Popperian falsificationism. And perhaps some philosophers are attracted by what certainly attracts Daniel Brooks (1983), one of the biological leaders of this approach. By turning away from notions of struggle and selection, one can break with the underlying themes of violence and selfishness which have (supposedly) been part of evolutionary theorizing since Darwin drew on the right-wing politico-economic philosophy of the Reverend Thomas Robert Malthus.

There are other challenges and queries about Darwinism. Some are based on empirical, especially molecular, considerations (e.g., Dover, 1982; also, for general discussion Ayala, 1972b, 1974d, 1975a; Ayala et al., 1974). Others concentrate on what they take to be conceptual inadequacies. The supposed inability of Darwinism to deal with complexity is often mentioned (e.g., Goodwin, 1984; Saunders and Ho, 1984; Stent, 1985; Webster and Goodwin, 1982; Lovtrup, 1987). And then, finally, in this account of challenges to Darwinism, we tip off the edge of science altogether. As is well known, in recent years the very idea of evolution as fact has come under challenge from religious fundamentalists, so-called Creationists. They argue that everything in the early chapters of Genesis — six days of creation, six thousand years total timespan, miraculous appearance of organisms, humans last, universal flood — can be justified by theories and principles of the best kind of science (Gish, 1972; Morris, 1974). I will not stay here to examine these claims for, as the world knows, they have been put together, in the form in which they are presented, simply as a ploy to get around the

U.S. Constitution's First Amendment separation of church and state (Montagu, 1984). Creation science is religion masquerading as science to get into the biology classrooms. It is interesting to note, however, that the Creationists have turned to philosophy to bolster their case. They are among the most ardent supporters of Popper, forever quoting his claims about the "metaphysical" nature of Darwinian theory. Since Creationists push this argument to the limit, claiming that any theory of origins is bound to be metaphysical, one does rather wonder at the strength of their other claim, namely, that their own views rightly belong in the science classroom.

Rather more significant is the fact that philosophers have stood behind biologists in defending evolutionary thought, both in writings and in the witness box, for the American Civil Liberties Union in its successful fight against a pro-Creationism law in the state of Arkansas in 1981 (Ruse, 1982a,b,c, 1984a; Futuyma, 1983; Kitcher, 1983). One interesting outcome has been a disagreement among philosophers themselves as to the proper way to distinguish between evolutionary ideas and Creationism ideas. Following the philosophical witness's advice (mine), the judge in the Arkansas case condemned Creationism as pseudo-science because (among other things) it is unfalsifiable (Overton, 1982). Several philosophers have taken issue with this conclusion, arguing that it is possible to protect any belief — including evolutionary belief — from falsification if one is so minded (Laudan, 1982; Quinn, 1984; Burian, 1986a). I have already rehearsed some of the counterarguments — about the ways in which one can generate genuine falsifiable predictions from Darwinian evolutionary theory. Here I will simply note that the defense of the judge's ruling has stressed the difference between being able logically to protect something from refutation at any cost, and its being reasonable to go on protecting something from refutation whatever the logic. The defense claims that the Creationists succeed only because they protect at any cost. (These questions are aired in detail in Ruse, 1988a. See also McMullin, 1986, for further background.)

With Creationism, we have gone as far as one can from Darwinism. No doubt there have been other ideas and challenges to Darwinism, and philosophers surely have or will take them up. But I trust that we have now a fair picture of the fervent goings on in evolutionary studies, and the ways in which philosophical ideas are intimately involved. Let me move on now to some other issues which have much engaged today's philosophers of biology.

CHAPTER 5

TELEOLOGY

A major philosophical puzzle about biology stems from the way in which its practitioners, unlike physicists and chemists, seem generally quite prepared to use teleological or functional language (Woodfield, 1976; Wright, 1976; Rescher, 1986; O'Grady, 1986). They talk about "ends" or "functions" or "purposes" — as in "the sail on the back of dimetrodon exists *in order to* regulate body temperature" or in "heartbeats serve no useful *function.*" Admittedly, every now and then biologists lecture themselves sternly on their practices. They assure themselves — and us — that their usage is just shorthand. And they try to pretend that they are not doing what they do, by refusing to talk of "teleology" and substituting "teleonomy" or some such thing (Mayr, 1974). But the language does persist. In the nineteenth century, that ardent Christian David Brewster (1854) assured us all that the moon exists in order to light the way of nocturnal travelers. No one today would ask about the function of the moon. But biologists do ask about the function of the kidneys or the purpose of the liver.

In the good old days, that is, in the days of Aristotle and his Christianized followers, *teleology* used to refer to life forces moving toward their goals, or to a Superior Being's plans and to the world being directed toward Its ends, or (in a pinch) to causes somehow working backward out of the future (Mayr, 1982; Losee, 1972; Friedman, 1986). As committed ontological reductionists, today's biologists do not saddle themselves with beliefs like these — at least, not qua biologists. There are, for instance, great difficulties with backward-working causes, like the problem of the missing goal object. (How does one explain a situation, if the goal is never achieved? See Nagel, 1961; Mackie, 1966; Ruse, 1971a, 1973a.) One has therefore to find some other analysis of teleology — an analysis which avoids the metaphysical commitments of the past and yet which still pays attention to the forward-looking nature of biology. (If one feels uncomfortable with *teleology* then by all means adopt the alternative, *teleonomy*. However, good biologists continue to speak of teleology, and my own feeling is to do likewise. After all, terms evolve, just like organisms. Do we mean the same by *atom* as the atomists?)

One of the most promising attacks on the problem of teleology came out of the Second World War, particularly by those people who had been working on or were impressed by machines (like torpedoes) designed to fix on a goal and to persist toward this target, despite external disruptions (Nagel, 1961, 1977; Sommerhoff, 1950; Rosenblueth et al., 1943; Falk, 1981). Surely, it was felt, we have here (in "goal-directed" or "directively organized" systems) a kind of nonmetaphysical teleology. Anything further from metaphysics than a torpedo it would be hard to imagine. Is this man-made machine not a model for biology? After all, directive organization is a widespread phenomenon in the living world. Think, for instance, of sweating and shivering, which "serve the end" of keeping the body at a constant temperature.

The claim was made, therefore, that biological teleology occurs insofar as — and teleological language is appropriate because — living things are goal-directed systems. They pursue ends, ultimately survival and reproduction, despite disruptions (not despite disruptions beyond certain limits, of course). Inorganic objects are not goal directed, and hence teleological language is inappropriate. That there may perhaps be borderline or anomalous examples (thunderstorms?) was thought to be a strength, rather than a weakness, of the analysis.

However, there are problems with this position. Although an organism may well be goal directed, that which leads us to speak teleologically may seem to have little to do with goal-directedness per se. The polar bear's white coat serves the end of camouflage, but there is no reason to think that, were the snow to melt, the coat could respond and get back on track toward camouflage by turning brown. As the late C.H. Waddington (1957) pointed out, talk of goal-directedness seems to point us towards *adaptability*, whereas when we are thinking teleologically about a feature like the polar bear's coat, we are concentrating on an *adaptation*. (See also Wimsatt, 1971.)

Perhaps this is the clue. Certainly, a number of people — philosophers and biologists — have decided so (Williams, 1966; Wimsatt, 1972, 1976b; Ruse, 1971a, 1972c, 1973a,b,c; Brandon, 1981). They unpack the teleology of biology in terms of adaptive advantage. When one says that something has a function, or purpose, or end, one is drawing attention to an adaptation: that is, one is referring to something that has been put in place by natural selection. The teleology comes about because one is trying to understand a phenomenon, say, the polar bear's white coat, in terms of the end it is supposed to serve, namely, camouflage. As with all adaptations, ultimately the end will be survival and reproduction. There is nothing suspect here, in the sense of implicit theism or backwards causation. Natural selection is a blind, materialist mechanism. The coat was caused by selection in the past, although, in trying to understand it today, we make reference to what we expect it *will* do.

Even if one accepts such a position as this, an appeal to natural selection — and no more — is incomplete (Ruse, 1981c, 1986a). On the one hand, one has more philosophical questions about the logical structure of such explanations and whether they can be fitted into a broader context of explanations covering also other areas of inquiry, especially the social sciences. The work of Charles Taylor (1964) has been very influential here. (See Achinstein, 1977; Baublys, 1975; Cummins, 1975; Jacobs, 1986; Utz, 1977; Nissen, 1983; Williams, 1976; Wright, 1972, 1973, 1974, 1976; Matthen and Levy, 1984, 1986; Bigelow and Pargetter, 1987; Matthen, 1988; Rosenberg, 1982b, 1986c, Bechtel, 1986c.)

On the other hand, one has more biological questions, especially the question, *Why* should teleology be so important in biology and not in physics? Grant that it is connected to natural selection. Why is natural selection a teleology-producing process, unlike, say, Newtonian gravitation? (This is a more than pertinent question, for there is no doubt that Darwin saw selection as a force entirely akin to gravitation.) Perhaps the clue to this latter question lies in the continuity between natural theologians like Archdeacon Paley (1802) and people like Darwin (who explicitly acknowledged the great influence that Paley had on his work). Of course, Darwin dropped the God-thought and all of that. (Actually, he did not do so quite so quickly as many people assume [Ruse, 1979a].) But, in important respects, the way of thinking and the language were identical. This fact makes one think that the natural theologians spotted something about the living world that today's biologists would acknowledge and exploit, through natural selection. And, of course, they did. They saw that the living world is *as if* it were designed. It is, metaphorically, like an object of human design, and (in the theologians' opinion) literally an object of devine design. The eye is like a telescope — to use Paley's famed example — because it is a product of design — metaphorically to us and literally to God.

What a number have suggested is that this "artifact model" is carried right over into evolutionary theory (Ruse, 1977b; Ayala, 1968, 1970a; Mayr, 1974; Lewontin, 1978). The Darwinian sees that organisms look as if they were designed. The heart is like a pump. The eye is like a telescope. The Darwinian knows that the effects come about through natural selection. There is no need to appeal, as once, to an omniscient god. But this does not deny the fact that living beings are like artifacts, objects of thought and planning. And because teleology is appropriate for artifacts

— they are made with ends in sight — teleology is appropriate for organisms (Williams, 1966).

If this is all so, then it might perhaps be suggested that those who located teleology in goal-directedness were not quite so misguided after all. If the teleology of adaptation rests ultimately in human ends — What is the point of the artifact you have just made? — then it is worth asking whether there is only one kind of human teleology. The answer is that there are more. There is not merely the teleology of created objects. There is the teleology of ourselves, as we strive to achieve goals — perhaps despite obstacles. I have the aim or goal of finishing my Ph.D., despite barriers like comprehensives and absent supervisors. I show that I am directed toward this end, precisely because I persevere, despite these roadblocks and like disruptions.

This thought suggests that goal-directedness, perhaps in its biological guise as adaptability, does in fact have a teleological dimension over and above that of mere function, namely, that which expresses itself in biology as adaptation. This is simply a suggestion. It is based on analogy or metaphor, and the big thing about analogies and metaphors is that some people will push them further than will others (Lakoff and Johnson, 1980). One may feel disinclined to allow sweating and shivering a teleological dimension above their adaptive function. On the other hand, even skeptics might feel ready to grant that conscious nonhuman beings, like chimpanzees, transcend the teleology of mere adaptation as they plot and scheme to achieve their ends (de Waal, 1982; Goodall, 1986).

Two more questions must be asked. First, is the teleology of biology (staying now with adaptation) a good thing? Or is it a sign of weakness and a snare into bad science? Of course, it will not have escaped anyone's attention that after a section dealing with challenges to Darwinism and worries about natural selection, we have rather slipped right back into focusing on Darwin's mechanism. Whether or not one personally believes that adaptation is all that there is to function, the point is that it is Darwinians who make adaptation central. Hence, if there is a tight connection between adaptation and function, we should expect to find Darwinians enthusiastic (or, at least, warm) toward teleology, and critics somewhat dubious. And this seems to be the case.

For instance, ardent Darwinians like the Harvard entomologist E. O. Wilson exploit the notion of function — especially as it can be used in devising so-called optimality models, by which one builds and tests systems that assume that one is using energy and resources as efficiently (optimally) as possible (Oster and Wilson, 1978; see also Beatty, 1980; Kitcher, 1987b; Sober, 1987a; and other contributions to Dupré, 1987). The very assumption here is that natural selection is simulating a good engineer who is trying to turn the cost/benefit ratio to his or her best advantage. Those critical of Darwinism tend to be wary of such approaches, and there is no doubt that at least part of the objection is rooted in discomfort at the quasi-natural theological nature of Darwinism. It is felt that this all smacks too much of a Christian age gone by (Gould and Lewontin, 1979). In particular, it is felt that it is inappropriate to assume that natural mechanisms will necessarily, or even generally, echo the planning and work of a conscious being. Such an approach is altogether too anthropomorphic. In the last chapter, we encountered some of the forms. that the objections take — for instance, accusing Darwinian teleologists of "just so" stories and unacceptable Panglossianism. If, to the contrary, one sees things like constraints as crucial causal factors in the evolutionary process, then the place for adaptation declines and the urge to think teleologically is reduced (although, see Caplan, 1981b).

Finally, could we get rid of teleology? One supposes that even Darwinians admit that in principle we could. One could talk always in terms of the past and of what selection did. But one would have to drop the artifact metaphor, and with it would go its incredible heuristic power (Beckner, 1969). Faced with a new organism, speculation would be barred. One would have to look simply at its performance in the wild. This seems an unduly harsh restriction in order to make a philosophical point. Apart from anything else, paleontology would become quite impossible. And in any case, linguists are always pointing to the broadspread use of metaphor right through our lives, including (especially including) physics. Why should biology be uniquely purified and crippled?

What implications does biological teleology have for questions of reductionism? A number of thinkers have argued that because biology is teleological in a way that physics is not, the possibility of complete reductionism is denied, at least at the methodological and theoretical levels. Biology will forever remain autonomous. (Interestingly, since these teleologists tend to be ultra-Darwinians, they also tend to be the same people who deny the impossibility of most reductionism from higher to lower levels *within* biology. See Ayala, 1968, 1970a; Dobzhansky et al. , 1977; Mayr, 1974.) However, there are those who argue that perhaps matters are a little more subtle than this. They point to the fact that where the physical sciences do meet the biological sciences, at the genetics interface, it is in fact the physical scientists who give way before teleology. Molecular biologists talk in functional language of the genetic "code" and of their success in "cracking" it. The point is that complex things which work do look like artifacts, even down at the molecular level, and thus teleological language is felt to be appropriate (Ruse, 1973a, 1977b; Rosenberg, 1985a,b).

What does this all mean? Certainly, so it would be argued, teleology bars the possibility of a straightforward reduction (methodological or theoretical) of biology to physics. However, where the two meet, perhaps it is the physical sciences which change in becoming teleological. And this means that, at least with respect to the theories or subtheories at the borders, talk of methodological and theoretical reduction is again made possible (although this is said without prejudice to the various points made earlier about the overall possibilities of such reduction).

CHAPTER 6

SYSTEMATICS

We come now to an area of biology which is inextricably bound up with philosophical questions, about which people have deeply differing intuitions, and in which the scientists (no less than the philosophers) inherit 2,500 years of conflict. I refer, of course, to systematics, the science of biological classification (Mayr, 1982). If we are going to do anything at all with the living world, we have simply got to make it manageable. We have got to sort organisms into groups or some such things, and then relate them to each other. Otherwise, we will be forever bogged down in talking about the particular. Fortunately, the job is not impossible, even though it sometimes seems this way to the worker in the field. Organisms do fall into groups or some such things, of their own accord, more or less. And the groups or such things do seem to be related, more or less.

Darwin (1859) put his finger precisely on matters when he explained that the relations of organisms, living and dead, are due to evolution (Beatty, 1985a). Unfortunately, it is there that the easy part ends. What are the groups? Are they really groups? What are the relations? Are they really relations? What is a given? What has to be interpreted? The questions flow on an on. I will here highlight the the two main questions for discussion, rigorously avoiding the millions of interesting side issues.

First, there is the "species problem" (Mayr, 1957). Although today this problem is couched in evolutionary terms, the controversy it causes raged before the *Origin*. It rages today no less violently (Mayr, 1982, 1983, 1987; Rieppel, 1988; Kitcher, 1984b, 1987a; Pratt, 1972; Ruse, 1969a, 1971c, 1973a, 1987b; Ghiselin, 1981, 1987; Rosenberg, 1980b,1985a, 1987; Sober, 1980, 1984a,c; Stebbins, 1987; Hull, 1970a, 1974a; Holsinger, 1984, 1987: Giray, 1976; Fales, 1982; Bock, 1986; Paterson, 1978, 1980; van der Steen and Voorzanger, 1986). The problem starts with the fact that, at some fairly basic level, organisms seem to sort themselves into groups, each organism belonging to one and only one such group. (In speaking of a problem, I speak philosophically. The existence of such groups is a matter of intensive relief to the worker in the field.) These basic groups are called "species," and what makes them interesting is the almost universal conviction that they are not arbitrary. Species are "real" or "natural" or "objective." I am a member of *Homo sapiens* and this really tells us something about me. By contrast, it is extremely useful to know that A comes before B in the alphabet, but a cataloguing of books using the alphabet hardly tells us something real about the books themselves. (Although see Grant, 1981, on botany).

Traditionally, there are two ways of dealing with the problem of groups, or (as it is known in philosophy) with the problem of "natural kinds" (Ayers, 1981). There is the Aristotelian approach, which looks upon entities as having essential attributes ("properties"), thus sorting such entities into real, objective sets. These essential attributes are to be distinguished from contingently possessed features ("accidents"). Although it may be possible to classify using just accidents, this approach yields only a "nominal" as opposed to a "real" definition. It just so happens that all and only humans are featherless bipeds, but this is merely a nominal definition. The real definition of humans is as rational animals. (For the pre-Aristotelian, Platonic view, see Kitts, 1987, although see Mayr, 1988.) Alternatively , we have the approach of John Locke (1959), which relegates all division to the nominal level, in Aristotelian terms. (Locke had a real-nominal distinction, but it was done in terms of the physical level at which entities are grouped.) Thus, for instance, any difference between men and changelings "is only known to us, by their agreement, or disagreement with the complex *Idea* that the name *Man* stands for" (III, vi, 39).

For fairly obvious reasons, as they stand, neither of these traditional approaches seems to do justice to the species problem. Biologists have a number of different ways of characterizing the groups which they would call "species" (Mishler and Donoghue, 1982; Kitcher, 1984b; Ruse, 1969a). Most obviously, it would be done in terms of some sort of morphological similarity. Darwin (1859) himself characterized a species as "a set of individuals closely resembling each other" (p.52). Today, one might attempt an analogous definition in terms of genetic similarity. Alternatively, one might refer to interbreeding and reproductive isolation (Mayr, 1969a). This is the basis of Ernst Mayr's (1942) "biological species concept." Or, if one is a paleontologist, one might be attracted to some kind of notion that pays respect to organisms as evolving lineages (Simpson, 1961).

Unfortunately, however one characterizes species, Aristotelianism is too strict. There are all kinds of variations within groups and loose ends between groups, barring a classification that crucially distinguishes between properties and accidents (Hull, 1965, 1974b). On the other hand, the Lockean answer is too weak. However characterized, species are undoubtedly Lockean natural kinds. But one wants to claim that the groupings are more than mere agreements with mental ideas.

At this point, two options face the philosopher. Either one aims for some other notion of natural kind, weaker than Aristotelianism yet stronger than Lockeanism, and applicable to species, or one entirely drops the idea that species are natural kinds, one comes up with some other characterization of them, and one shows that this is why we look upon species as natural. Both options have been taken, by philosophers and by biologists. As we shall see, some of the motives lie in the deep disagreements discussed already.

Those who would argue that species are indeed natural kinds usually begin by allowing variation within groups (Beckner, 1959). Instead of demanding attributes which must be possessed by all members of some specified group, it is conceded that groups might be characterized by a "cluster" of attributes. Species can thus be given "polythetic" or "polytypic" definitions, whereby membership is determined by possession of some from a set of attributes, no one of which is necessary but a number of which is sufficient. This, of course, still leaves open the question of why one should think that species are natural or objective. A move favored by a number of thinkers makes a virtue of the various ways in which species are characterized (Ruse, 1969a, 1973a, 1987b; Mishler, and Donoghue, 1982; Mishler and Brandon, 1987; Dupré, 1981; Kitcher, 1984b,c, 1988). It takes heed of a principle enunciated by the nineteenth-century thinker William Whewell (1840), who had the distinction, not only of being a professor of philosophy, but also of mineralogy (where classification is as important as to the biologist). He wrote:

> The Maxim by which all Systems professing to be natural must be tested in this: — that the *arrangement obtained from one set of characters coincides with the arrangement obtained from another set.* (Whewell, [1840], I, 521, his italics. For more details on how Whewell used his ideas in mineralogy, see Ruse 1976b.)

Whewell's idea here is that, if quite different features pick out the same set, then one knows that one is on to something, but not otherwise. An alphabetical classification of books would never have come out that way had the alphabet not been used — whereas all and only those substances with the physical features of common salt are those substances with the chemical formula $NaCl$. Likewise, it is argued that a classification of, say, birds solely on grounds of age would find no correspondence with other ways of making groups. Against this, a classification of birds based on general morphological similarities can also be achieved quite independently of morphology — through genetic similarities or interbreeding, for instance (Ayala, 1974a, 1975b). Thus, this is why we think that the group we call *Passer domesticus* is natural, whereas the class of birds aged exactly one year, two months, and five days is not. (For other thoughts on the naturalness of species, see Caplan, 1980, 1981a; Caplan and Bock, 1988; Kitts and Kitts, 1979.)

The other option is to drop the hope entirely (or, if one likes, the pretence) that species are natural kinds (Rieppel, 1986). This is the suggestion of the biologist Michael Ghiselin, vigorously and impressively supported by the philosopher David Hull (Ghiselin, 1966, 1974, 1981, 1987; Hull, 1976a, 1978a, 1980; also, Rosenberg, 1980b 1985a; Sober, 1980, 1984c; Williams, M.B., 1985, 1987, 1988; Beatty, 1982a; Mayr, 1976; Crowe, 1987; Cracraft, 1987). They argue that species are not classes at all. They are *individuals*. What does this mean? To a certain extent it means what we make of it — but we do have some paradigmatic examples of individuals, namely, organisms. Hence, in a sense — in the vitally important sense that this is the way evolutionary biology treats them — species are like supraorganisms. They have an integration, an internal organization, which binds them into a unified whole, quite unlike a normal class or natural kind.

This is a strong claim, but is backed by forceful arguments. Ghiselin, Hull, and supporters turn to evolutionists like Mayr (1969a), finding that they claim explicitly that species are individual-like. "Species are the real units of evolution, they are the entities which specialize, which become adapted, or which shift their adaptation" (Hull[1976a], p.183 quoting Mayr [1969a]). They find that, from a temporal perspective also, evolutionists treat species as units. A class cannot change over time. Species evolve and change over time — as, in analogous manner, do individual organisms. And, in the way that one specimen serves as the marker for a species — it is the "type specimen" — Ghiselin and Hull find support for their view.

Discussion continues, and as with most philosophical controversies, both positions have strengths and weaknesses.

There really is a great deal of variation within many species. Whether one can subsume it completely beneath a polytypic definition is often more a matter of faith than of proven fact — as also is the confident assertion that a species delimited one way (e.g., genetically) will correspond with a species delimited another way (e.g., morphologically). However, on the other side, the species-as-individuals thesis has some odd, not to say paradoxical, consequences. If an individual organism dies, then barring bodily resurrection, that is it. It can never come again. But would we want to say the same of a species? Suppose a new organism is produces through polyploidy (the combination of complete sets of chromosomes from the hybridization of individuals of different species). Suppose then that all the members of this new species are destroyed, and then at some later point new, similiar organisms are produced. Surely we have new members of the same species, not a new species?

What some Darwinians find particularly troublesome about the species-as-individuals thesis is that it seems to go flatly against the renewed biological emphasis on individual selection (Ruse, 1987b). By stressing the unity of species one is downplaying the fact that every organism — perhaps every gene — is (in a sense) set against every other organism or gene. There may be cooperation (we shall see later that there is), but it must rebound to the benefit of the individual. To think of the species itself as an individual is to ignore the internally competing interests. Perhaps this point compels. Perhaps it does not. But it surely suggests that it may be futile to look for the one unique solution to the species problem — at least, a solution which is apart from the rest of one's thinking. One's stand on the species-as-individuals thesis will be very much bound up with one's attitude toward reductionism, especially methodological reductionism. If one looks to explain the larger in terms of the smaller, one will feel uncomfortable with the thesis. If one believes that larger entities have properties in their own right, one will incline towards the thesis.

Hull particularly insists that the debate cannot be resolved by common sense, but demands reference to actual biological thought. This idea meshes with what has just been suggested in the last paragraph, and we do in fact find that those scientists and philosophers who have most strongly embraced the species-as-individuals thesis are precisely those trying to go beyond ultra-Darwinism to less reductionistic views of the evolutionary process. The thesis is an integral part of modern hierarchical thinking, and is the cornerstone of species selection; truly, it is only by thinking of the species as an individual that one can properly think of one species being selected rather than another. And again, the thesis lends itself readily to punctuationist thinking, for there it is supposed that species — like individual organisms — have fairly distinct beginnings and ends. All of this inclines one to think that the same paradigm difference enters as specifically into taxonomic thought as it does into general evolutionary thought. (See, especially, Gould, 1982a; Eldredge, 1985a,b; Brooks and Wiley, 1986; Grene, 1987.)

The second major philosophical problem thrown up by systematics concerns the correct way to classify, looking now at organisms from the broader overall perspective (Buck and Hull, 1966; Hull, 1970a, 1978b; Mayr, 1981; Ridley, 1985; Platnick, 1982; Platnick and Nelson, 1981; Snyder, 1983; Sober, 1988a). Suppose one has placed organisms more or less into their respective species. Where does one go from here? One wants more than a simple catalogue, with *Homo sapiens* coming after *Canis familiaris* but before *Tyrannosaurus rex*. But how much more does one want and how much more can one have? Essentially, in answering these questions, there seem to be two extremes, with (as the biologist Mark Ridley has well remarked) most people coming somewhere in the middle.

At the one extreme, there are those who favor a fairly direct evaluation on the basis of physical properties, with classification being done in terms of similarity and difference (Sneath and Sokal, 1973). One looks at an organism, breaks it down into characters, and then, compares it with other organisms similarily broken down. If *Homo sapiens* comes out as more similar to *Canis familiaris* than to *Tyrannosaurus rex*, then that is the basis for classification, but not otherwise. On the basis of repeated comparisions, one can build up an overall ordering of organisms. (As noted above, this ordering would usually be of organisms, as already collected into species.)

Although followers of this school, "numerical or phenetic taxonomy," are committed evolutionists, they want to keep their evolutionizing out of their day-to-day work. Evolutionary speculation is to be done on the basis of a classification, not fed in as a guide to the classification. And the justification offered is explicitly philosophical. Supposedly, only the kind of empirical nontheoretical approach of the phenetical taxonomist guarantees objectivity. In any case, were one to use evolution as a basis of one's classification, one could not then use the classification in the quest for evolutionary understanding — at least, one could not do so without the risk of vicious circularity.

At the other extreme, we have the classifiers for whom evolution is all — at least, that was the way that things started out. They argued that phylogeny is the ultimate basis of any classification (Wiley, 1981; Cracraft, 1974, 1978; Eldredge and Cracraft, 1980; Janvier, 1984; see also Hull, 1979c, 1981b, 1984c, 1985a, 1988). If *Homo sapiens* is in some way more closely related to *Canis familiaris* than to *Tyrannosaurus rex*, then classification should reveal this fact — even if, physically, *Homo sapiens* is far more like *Tyrannosaurus rex* than *Canis familiaris*. This school of classification, known (now rather more rarely) after its founder as "Hennigean" (Hennig, 1966) or (now more commonly) as "phylogenetic" or "cladistic," seems to break down into two main components, with supporters insisting on both. First, there is the actual way in which one is supposed to construct a classification. Second, there is the method one is supposed to use as one aims towards such a construction.

The first component is probably more controversial — at least, less readily accepted by nonpartisans. In the name of objectivity (again a philosophical appeal), it is insisted that points of branching should be the key to classification. If there is no branching, then no new groups can be formed. If there is branching, then new groups must be formed. All organisms which are descendant from members of a group just after branching must at that level be included in that group. There is no choice, and hence no unwanted "subjectivity."

But how does one know the true phylogeny and the required points of branching? Here enters the second part of cladism, the method — something which is probably appreciated by many taxonomists who feel uncomfortable with the first component of cladism. Essentially, one is instructed to distinguish between "primitive" and "derived" characters, the former going back before a point of branching and the latter coming after. In the light of such distinctions, one is thus able to distinguish among crucial aspects of phylogeny. But how is one to distinguish between primitive and derived? Here the aim is to throw up alternative phylogenies and, judging on the empirical evidence, to choose that (or those) which require(s) the least appeal to extraordinary or unusual evolutionary processes, like extreme convergence. It is assumed that it is more likely that similar features are due to shared ancestry than to all the complications of independent evolution (Ridley, 1985).

Much appeal is made at this point to the virtues of simplicity or parsimony. Again philosophy rears its head, for, linking simplicity with openness to falsifiability, cladists have rather adopted Popper as their patron saint. ("Simple statements, if knowledge is our object, are to be prized more highly than less simple ones *because they tell us more; because their empirical content is greater; and because they are better testable"*: Popper, 1959, his italics [Wiley, 1975, 1981; Nelson, 1973; Nelson and Platnick, 1981; Eldredge and Cracraft, 1980; Gaffney, 1979].)

It is possible that most taxonomists try to steer a course between the extremes, taking note of phylogeny but being sensitive also to overall similarities. With Ernst Mayr (1969a, 1981), they think this to be the correct "evolutionary" approach. But decisions about approaches are decisions for biologists. What of decisions for philosophers? Since the taxonomic extremes are based so greatly on philosophical claims, about objectivity and falsifiability and the like, what philosophical queries (support or objection) have been raised about them?

In the case of phenetic taxonomy, the responses have been almost entirely negative, pointing out that any such supposed non-theory-laden objective classification is almost certainly a sham — quite apart from the purely empirical question of whether independent observers would ever come up with the same classification. As soon as one starts breaking organisms into parts, one must bring in theory (Hull, 1968; Ruse, 1973a). Take two bears, one white and one brown. Do they differ in one feature, or does one take each hair separately and talk of literally hundreds of thousands of differences? (Of course, I know what the answer is. The point is whether someone who explicitly eschews the theory has the right to combine all of the hairs into one feature.) This is not to say that the very enterprise of identifying parts and trying to assess relationships is necessarily flawed (Ghiselin, 1966); in fact, it is usually an essential part of taxonomic practice. But, it is to say that the philosophical claims about non-theory-laden objectivity are illusory. (Against the circularity worries, see Hull, 1967.)

At the other extreme, the cladistic way (as opposed to technique) of classifying divides philosophers as it divides biologists — and for some of the sorts of reasons which, by now, must be starting to sound very familiar. Let us agree that if one bases division purely on points of branching, then one has an automatic repeatable method of taxonomy. But is all significant evolutionary change — all change that one would want to record in a classification — captured by reference to branching? If one is the kind of evolutionist who thinks that change generally comes at the time of the formation of new species, then one probably thinks that the answer is mainly positive. However, Darwinians would see gradual change without branching ("phyletic gradualism"), which they would want to record (Gingerich, 1976, 1977). Should *Homo habilis* be put with *Homo erectus* be put with *Homo sapiens,* despite the fact that the first has a brain size of 500 ccs, and the last a brain size of 1,400 ccs? Most would feel uncomfortable with such a grouping, feeling at this point that the virtues of cladism are purchased at too high a price.

Of course, if one thinks that there was branching when *Homo erectus* and *Homo sapiens* evolved, then as a cladist one can separate them. But the orthodox Darwinian doubts such branching. It is surely no surprise that many of the leading cladists are people who are stretching Darwinism, most particularly Niles Eldredge (1985a,b). The same goes, even more so, for those, like

E.O. Wiley, who really wants to break away from Darwinism; given his thermodynamical commitments, he particularly (with Brooks) thinks that points of speciation are *the* points of evolutionary action (Brooks and Wiley, 1984, 1986). And it is no surprise either that those philosophers who feel comfortable with cladism as an overall approach are precisely those who want to downplay adaptationism and play up other factors in the evolutionary process. (For many, the species-as-individuals thesis and cladism go hand in hand. If one thinks that a species is an individual with a clear beginning and end, there are obvious attractions in a taxonomic method which stipulates that beginnings and ends must be clear-cut. Moreover, things like the unique historical nature of the species as individual might be thought to lend themselves to cladism. On the other hand, one cannot make too much of possible links. Gould accepts punctuated equilibria theory, the species-as- individuals thesis, but rejects cladism as a complete philosophy of classification. See Gould (1982a) for acceptance of the first two points and Gould (1980b) for rejection of the third.)

What of the techniques of the cladists, working from the distinction between ancestral and derived features? As noted, much is made of the Popperian virtues of this technique, with appeals to the way in which assumptions about the independent evolution of features ("homoplasy") falsify presumed phylogenies. Given the general philosophical difficulties with the Popperian appeal to falsifiability — most particularly, that one can always shore up a hypothesis in the face of counterevidence — many philosophers incline to give the whole procedure short shrift. This is surely too extreme a reaction. If one thinks of falsifiability in a rather weaker sense — the best hypothesis is the one which requires the least improbable supports (e.g., the least number of appeals to homoplasy) — then the technique is powerful. Moreover, this conclusion is backed by reasons along Popperian lines (Sober, 1983b, 1988a; Felsenstein and Sober, 1986).

However, it must be noted that the appeal to Popper has (in the opinion of some) backfired in one rather unfortunate respect. Whether because of Popper's specific difficulties with Darwinism per se — difficulties which, for reasons mentioned above, are shared by many cladists — or whether because of what is taken to be a rigorous application of general Popperian principles, there are those cladists who have gone full circle, and from being the ultimate phylogeneticists have become nonphylogeneticists. They argue that, since we are working with the features of specimens laid before us, we should make no inferences about the past. Classification refers only (and should refer only) to the characters of organisms and says (and should say) nothing as such about phylogenies. To say more goes beyond the simplest, most falsifiable hypothesis (Patterson, 1978a,b, 1981, 1983).

In the opinions of critics, this subschool — "pattern" or "transformed" cladism — seems to suffer from the same philosophical blinkers as did Woodger. Critics argue that good science — philosophically acceptable science — is not science shackled by a crude empiricism, which can refer to nothing but the seen-here-now. It is science which makes all kinds of claims about the unseen and unknowable — even the unseen and unknowable in principle. What good science demands is that all of its claims be linked in some fashion to the world of sensation. But this is a very different demand from that allowed by the pattern cladists. Critics conclude that a taxonomy today which does not make reference to the past is a taxonomy unduly truncated (Beatty, 1982a; and replies by Patterson, 1982; Platnick, 1982; and Brady, 1982; also see Ridley, 1986). (In fairness to Popper, I must emphasize that his own philosophy goes far beyond a crude empiricism.)

CHAPTER 7

HUMAN
BIOLOGY

Humans may not be the most important organisms. If Darwin was right, any such talk of this nature is suspect. However, they are the most interesting — to us, that is. They are also the most threatening, in the sense that a biological account of humankind is not simply a disinterested exercise in scientific understanding. It is something that may well challenge some of our deepest and most cherished and comforting beliefs and prejudices (Ayala, 1973, 1986). This, no doubt, is a major reason why there is so much tension around philosophical discussions of the biology of humans, and also why such discussions are wont to break out into sheer nastiness. If we really are the end results of a blind process of evolution, then this fact matters — both now and probably in the long- term future. Thomas Henry Huxley and Bishop Wilberforce of Oxford fought a celebrated duel in 1860 over our nature at the British Association for the Advancement of Science (Lucas, 1979). Echoes of that clash can still be heard (Dobzhansky and Ayala, 1977).

At the strictly scientific level, the past twenty years have been incredibly exciting times for human biology. Thanks, particularly, to fossil discoveries and to newly developed molecular techniques, we now know a great deal about the nature and causes of human evolution (Issac, 1983; Johanson and Edey, 1981; Pilbeam, 1984). We know, for instance, what would have been beyond belief a mere ten years ago, that the human and chimpanzee lines split a mere 6 million years ago. For at least 3 1/2 billion years we were one. Only in the last instant of geological time have we broken away. We know also the answer to a problem that has plagued evolutionists ever since Darwin. Humans first got up on their hind legs. Only then did their brains start to develop.

Actually, however, although some commentators of science have been very much interested in the methods and assumptions behind work on human ancestry — one thinks here particularly of the sociologists and historians who have shown just how close to fiction so much supposedly sober scientific work really is (Landau, 1984; Bowler, 1986) — philosophers have been relatively indifferent to the past per se. Their interest lies much more in the biology of today's humans, particularly in the efforts of the "sociobiologists" to provide an understanding of human nature (Caplan, 1978a, 1984; Ruse, 1979b; Kitcher, 1985a, 1987c; Midgley, 1978). Here, it is felt, there are many issues which transcend the purely empirical, calling for detailed philosophical analysis. Let us see why.

As is well known, the past few years have seen a vigorous attempt by Darwinian evolutionists to push their theories into the realm of social behavior: to create the new discipline of "sociobiology" (Wilson, 1975; Dawkins, 1976; Bonner, 1980). It is argued that selection and adaptive advantage are as important in understanding what organisms do (particularly to each other) as they are important in understanding what organisms are. Most particularly, working from a strong individual selectionist stance (see above), Darwinians have proposed models and explanations showing why animals cooperate (are "altruistic": Hamilton, 1964a,b), why relations sometimes break down, bringing on aggression of various kinds (Maynard Smith, 1974, 1984b), why there are differences between the sexes and when and how mates cooperate or compete (van den Berghe, 1979), why some animals look after their offspring and others do not (Trivers, 1972), and why many other behaviors occur.

As is even better known, many of these Darwinian evolutionists have gone straight on to apply their models and explanations to humans. Thus, from the self-same individual selectionist stance, Darwinians have proposed models and explanations showing why humans cooperate (are "altruistic"), why relations sometimes break down, bringing on aggression of various kinds, why there are differences between the sexes and when and how mates cooperate or compete, why humans are religious, why some humans look after their offspring and others do not, and why many other human behaviors occur (Wilson, 1978; Alexander, 1976, 1979; Reynolds and Tanner, 1983; Barash, 1982; Hamilton, 1984). (There is absolutely no reason in principle why one should not have a non-Darwinian human sociobiology, but in practice it seems not to exist — except, perhaps, inasmuch as one might deem an occasional yearning for group selection to be non-Darwinian. See Sober, 1988b, for an interesting philosophical discussion of altruism that raises individual and group selection issues.)

Claims about altruism and the like — the very heart of human sociobiology — have been highly controversial, calling down derision and scorn from many, particularly left-wing biologists (who loathe any attempt to relate humankind directly to the genes) and social scientists (who feel that sociobiologists are insensitive to their concerns and who seem incidentally much threatened by the moves of sociobiologists into their own pitches). Much of the attention of the critics has, naturally devolved upon empirical questions (Allen et al., 1977; Lewontin, 1977; Lewontin et al., 1984; Sahlins, 1976; Kauffman, 1977). But philosophers — with very few exceptions — have likewise rushed to criticize and attack the subject. (For instance, see Flanagan, 1981; Burian,

1981; Kitcher, 1985a, 1987c; Sober, 1985a,b; Holcomb III, 1987; Grene, 1978; van der Steen and Voorzanger, 1984a,b; Voorzanger, 1984, 1987a,b. On the other side, see Ruse, 1977c, 1979b; Rosenberg, 1981, 1983c. A centerline position is found in Caplan, 1981b, 1984, and his edited collection, 1978a. See also Thompson, 1980; Shaner, 1987; Campbell, 1975, 1983, 1986; Hull, 1978b,c. For a short history, see Segerstrale, 1986.) Roughly speaking, the philosophers' objections have been of two kinds, which I will take briefly in turn.

First, there are questions to do with values, or what one might call "normative" matters. It is argued that human sociobiology, implicitly or (all-too-frequently) explicitly, denigrates certain groups of people: blacks, Jews, the lower classes, and, most particularly, women. Human sociobiology, supposedly, is committed to a thesis of "genetic determinism," whereby all peoples are locked blindly into their roles and statuses by their biology, with no possibility ever to changing things (Lewontin et al., 1984; Gould, 1981). Thus blacks and others are regarded as "different," a euphemism for "inferior." People who are less successful in life's challenges are thought to be condemned to misfortune by their genes. And women are forever relegated to the traditional roles of *Kinder, Kirche, Kuche.* Human sociobiology claims that they are, by nature, milder, softer, and less intelligent and motivated than men. To use a notorious term (of E.O. Wilson in his *Sociobiology: The New Synthesis),* women are more given to *coyness* than men.

Obviously, argue the philosophical critics, all of this has more to do with the deep-seated prejudices of white, Anglo-Saxon, middle-class, academic males than with anything in the real world. Values — obnoxious values — are being read into the record, and then read right back out again, as though they existed objectively. Then these bogus conclusions are used to prop up the status quo: the racist, capitalist, sexist status quo. What we should realize, rather, is that there is nothing in the world, certainly nothing in the world of biology, to support any of these beliefs (Ruse, 1985b).

Paralleling these criticisms from a value standpoint, we have also criticisms about the genuinely scientific nature of human sociobiology — what might be called "epistemological" criticisms. Here the charge is that human sociobiology fails to engage with the real world in any genuine way (Lewontin, 1977). It has all of the faults of traditional Darwinism, and then some. It gives pseudoexplanations, thinking up "just so" stories, to explain any and every human phenomenon from an "adaptationist standpoint." Panglossianism rears its head higher here than it does anywhere else (Gould and Lewontin, 1979). Take anything one likes to mention — no matter how far it may be removed from biology — and sociobiologists will come up with a selectionist scenario. Homosexuality is a paradigmatic example. Sociobiologists argue that it must be due to balanced superior heterozygote fitness or to kin selection (thus making homosexuals like sterile worker ants) or some such thing. No matter what, it will be related to reproduction, real or apparent (Gould, 1978; Kitcher, 1985a).

Obviously, argue the critics, this is no true science. What we have is mere gameplaying, a redescription of the phenomena in fancy terms. To put matters in Popperian language, human sociobiology is simply not falsifiable (that is, when it is not outrightly false). It tells us no more about ourselves than Christian Science tells us about disease. If we were to follow Popper in speaking of a "metaphysical research programme," we would be generous.

Philosophical defenders of human sociobiology freely admit that human sociobiology is a long way from a proven, solid science. Like any other would-be inquiry, true empirical evidence must be sought and found. In fact, defenders argue that such evidence is now starting to come in. However, this is not my direct concern here, for I will take only the philosophical objections and their possible responses.

Take first the question of values. All must admit that sociobiologists have not always been careful about either language or models. Undoubtedly, some fairly heavy-handed sexist comments (to name but one area) have occurred — although, how pure any of us were until about ten years ago might be questioned (Ruse, 1986c). On the other hand, argue defenders, this is a far cry from saying that the whole subject is irretrievably tainted, or even that already-existent careful formulations are tainted. Indeed, it might be maintained that, far from stressing differences between people, as much as anything sociobiology underlines similarities. It is claimed, for instance, that humans cooperate for much the same reasons in industrial societies as they do in preindustrial societies. Is this not to make us all brothers and sisters under (and over) the skin?

It is true that, to take the most contentious issue, sociobiologists pick out and discuss male-female differences, believing that these (including emotional differences) will not be rooted entirely in culture or the environment (Symons, 1979). But, argue defenders, there are no claims that females are inferior to males — certainly not in the human qualities that count, like sensitivity or intelligence. Indeed, in important respects sociobiology emphasizes that males and females must be balanced and equal. Natural selection ensures that females must be as much in control of their destiny as males. Otherwise, we might as well all be born males. (See also Hrdy, 1981, for more on this point. Those who claim that Darwinian evolutionary theory and other parts of biology are sexist include Birke, 1986; Bleier, 1984; Haraway, 1983; Harding, 1986; Harding and Hintikka, 1983; Hubbard, 1983; Keller, 1983, 1984, 1987; and Reed, 1975. Ruse, 1985b, 1988d, takes a less critical view.)

What of the charge of genetic determinism? It is certainly agreed that biology makes an input into our nature — that is the whole point of human sociobiology. But the response is that this is a far cry from saying that we are just blind robots controlled by the DNA (Ruse, 1987a; although see Kitcher, 1985a). To use another notorious phrase (also see Wilson, 1978), at most the sociobiologist wants to claim that "the twig is bent a little." The defender of the subject allows that our minds are not tabulae rasae — indeed, the sociobiologists would assert explicitly that there are innate dispositions which inform and guide our thought and action (Lumsden and Wilson, 1981, 1983). But how exactly these dispositions get filled up and made explicit is all very much a matter of how environmental (especially cultural) factors impinge on the growing and adult human — factors, of course, over which we ourselves have at least some control.

So much for the defence against the normative criticisms. A similar line is taken against the epistemological criticisms. Undoubtedly, some sociobiologists enthusiastically see adaptation where none exists — at least, where we have no right to say that any exists. But this is a far cry from saying that the whole subject is a nonscience (Caplan, 1981b). Moreover, we are urged not to forget that human sociobiology is but one part of the overall Darwinian synthesis. Hence, as a relatively new subject, it can legitimately draw on the rest of the theory for support in what it claims (Ruse, 1979b, 1982a, 1987c; although see Caplan, 1984; Thompson, 1985). If the sociobiologist centers his or her attention on, say, homosexuality — a natural enough move given that homosexual orientation seems to fly in the face of Darwinian goals — he or she is not transgressing into nonscience by seeking adaptive causes (Ruse, 1981a, 1988d). All the evidence we have of the biological world is that adaptation is crucial. Therefore, the legitimate presumption when dealing with new types of phenomena is that adaptation is important here also (Borgerhoff Mulder, 1987; Caro and Borgerhoff Mulder, 1986). Time may prove that it is not. But the presumption of importance is the mark of good science — not pseudoscience. (The exact place of human sociobiology with respect to the rest of Darwinian theory has, in fact, been a matter of recent debate, with some arguing that it fits right in, and with others arguing that it does introduce new principles: Caplan, 1984; Ruse, 1987c.)

These, then, are some of the moves taken by the philosophical defender of human sociobiology. Although I have myself been much involved in the controversy over the subject, as reviewer rather than as participant I make no overall assessment of the strength and worth of human sociobiology. It is enough that I have shown that a major element in the debate is more philosophical than strictly scientific. The place of humans in the natural order of things is as much a matter of philosophical concern as it is of scientific concern. But do let me note that the sociobiological controversy is not separate from controversies we have considered earlier. Although no doubt one could be a sociobiologist and a supporter of punctuated equilibria theory, as I have noted, the critics of human sociobiology (at the philosophical level particularly) tend to be those who would stretch or even bend neo-Darwinism. As common as the charge of "determinism" is the charge of "reductionism." The feeling is that the very attempt to understand human nature simply in terms of our constituent biological parts is a manifestation of the bad science which we have seen critiqued earlier. Good science demands a holistic, hierarchical approach, and this requirement applies as much to human nature as it does to the fossil record. (Paradoxically, given the charges, one finds that no one is more holistic in his regular science than is E.O. Wilson — see, for instance, Wilson, 1971, 1985.)

It would be easy to say, given the explicitly Marxist commitments of some of the leading critics of human sociobiology, that this is all simply a matter of politics; but perhaps it makes more sense to say ultimately that the opposition to ultra-Darwinism, whether in paleontology or human sociobiology, is more a reflection of the already-discussed, different, European way of viewing the evolutionary process. Human sociobiology certainly intensifies the debate; but what is being intensified is not altogether new or separate from other controversies. (For more on the human sociobiology question, generally critical, see Barlow and Silverberg, 1980; Leeds and Dusek, 1981; Montagu, 1980; Midgley, 1978; Schneewind, 1978: Searle, 1978; Washburn, 1978. Recent work on gene/culture coevolution includes Boyd and Richerson, 1985; Brandon, 1985d.)

CHAPTER 8

PHILOSOPHICAL IMPLICATIONS

Thus far, I have been surveying philosophical discussions about the nature of biology. The former subject has been brought to bear on the latter subject. Can the process be reversed? Should it be? Does biology have anything to tell us about the traditional problems of philosophy? In particular, can biology tell us anything about the theory of knowledge, about epistemology? Can biology tell us anything about the theory of morality, about ethics?

Many people — most philosophers — would answer negatively to both of these questions; that is, to anything but a trivial formulation. (Obviously, one must have some biological knowledge if for instance, one is going to talk about the moral desirability of genetic counseling.) It would be agreed that we have our powers of thought and reason, our abilities to perceive the world, and our capabilities for moral action, in part (probably in major part) because of our evolutionary past. But it would also be argued that none of this is relevant to the ways we think and act now. A truth of mathematics, like the Pythagorean theorem, is true independently of biology — as also is our appreciation of its truth (Nozick, 1981; Nagel, 1986). And the same is true of morality. Whether or not a good god stands behind the standards of right and wrong, morality and our appreciation of it is something above and beyond the forces of natural selection (Trigg, 1982; Singer, 1981; Peacocke, 1986).

There has always been a minority of dissenters — a minority which has demanded some attention, if only because it has contained some of the most distinguished of evolutionary biologists (Dobzhansky, 1962; Huxley, 1947; Waddington, 1960; Wilson, 1975, 1978). But in recent years this minority has been swelling, augmented by not a few good philosophers (most notably, Quine, 1969a,b; but see also Giere, 1985; Richards, 1977; von Schilcher and Tennant, 1984; Wuketits, 1986; Thompson, 1988b,c). Those who would argue that biology is important for philosophy are still on the outside, looking in. However, "evolutionary epistemology" and "evolutionary ethics" are now topics that any survey such as this must acknowledge and discuss. (Campbell, 1974c, and Campbell et al., 1986, contain excellent evolutionary epistemology bibliographies. And see also Hahlweg and Hooker, 1989; and Plotkin, 1982, 1987. Alexander, 1987, is a good review of evolutionary ethics.)

Taking first the attempts to bring biology (particularly evolutionary biology) to bear on epistemology, we find two main approaches (Bradie, 1986). First, there are those that work through *analogy*. It is noted that the growth of knowledge (particularly the growth of scientific knowledge) is continuous — it is evolutionary. Hence, it is suggested that perhaps the causes behind the growth of knowledge are like the causes behind the growth of organisms. Specifically, perhaps knowledge grows through a kind of selection process. And, it is concluded, this is indeed the case. Ideas are thrown up, there is a "struggle" for people's attention and allegiance, and then certain of these ideas are "selected" and kept until new challenges arise (Toulmin, 1967, 1972; Campbell, 1973, 1974a,c, 1977, 1987a,b, 1988a,b; see also Hull, 1982a,b, 1984a,b, 1986, 1987, 1988a,b; Griesemer and Wimsatt, 1988; Wimsatt, 1985a; Darden, 1987). In recent years, this position has received a powerful boost from Popper (1972, 1984), who claims that (in science, at least) the selection of ideas is simply his notion of falsification by another name.

The second approach to epistemology takes the biology *literally*. Here it is argued that natural selection has left the human mind informed by various innate dispositions (as suggested by the human sociobiologists), and these dispositions incorporate our principles of thought and reasoning (Popper, 1963; Ruse, 1985a, 1986c; Lorenz, 1941; Sober, 1981a,b; Giere, 1985; Tennant, 1983b,c, 1988a,b; Brandon and Hornstein, 1986; Lieberman, 1984; and historically, Smith, 1987). Thus, for instance, the Pythagorean theorem seems compelling to us not because it reflects an eternal verity, but because those of our would-be ancestors who took seriously Euclidean geometry outreproduced those of our would-be ancestors who were indifferent to the problems posed by space. This kind of evolutionary epistemologist certainly does not want to say that every last knowledge claim that we would make is rooted in biology. Ultimately, however, all that we believe and think is hung on a mental skeleton, put in place because it confers (or conferred) adaptive advantage on us humans.

Turning now to ethics, we likewise find two approaches (Tennant, 1983a). First, there are those who believe that the biological evolutionary process itself generates value, including moral value (Huxley, 1947; Wilson, 1978; Waddington, 1960). It is argued that, in some very real sense, evolution is progressive, stretching ultimately from monad to man (Ayala, 1972a, 1982). Hence, because the end product clearly has more worth than the beginning crude life forms, it is morally incumbent upon us at least not to hinder the evolutionary process, and perhaps even to cherish and positively aid it. How, exactly, this aiding and cherishing might be performed has been a matter of some interpretation. Best known — most contentiously — there is traditional social Darwinism, with beliefs about the virtues of struggle and a commitment to all-out unrestricted capitalism

(Russett, 1976; Jones, 1980). Other such evolutionary ethicists take a milder line. Andrew Carnegie, the steel magnate and philanthropist, funded public libraries so that the poor but deserving ("fit") child might rise in life. Edward O. Wilson (1984), an ardent conservationist, would have us protect wildlife because he believes we humans depend symbiotically on such life (although see Singer, 1986).

The second evolutionary ethical approach corresponds directly to the second evolutionary epistemological approach, taking Darwinian evolution literally and homing in on the innate dispositions which inform our thinking (Ruse, 1986c; Ruse and Wilson, 1986). It is argued that sociobiology in particular shows that cooperation can be a good biological strategy. In many circumstances, individuals get more out of life (improve their reproductive fitness) if they work together, than otherwise. But how does natural selection get us to cooperate, given that (for fairly obvious reasons) our natural dispositions are to be at least self-centered, if not outrightly selfish? Simply by incorporating morality within the innate dispositions! Just as we believe the Pythagorean theorem to be true, because it is in our biological interests to do so, so also we believe that we should love our neighbor as ourselves, because it is in our biological interests to do so. Morality has no more, and no less, function and status than any other biological adaptation (Mackie, 1978; Murphy, 1982; Gibbard, 1982). (We do not consciously choose to believe in Pythagoras or morality. We probably function better if we think them imposed on us by objective reality.)

Evolutionary epistemology and evolutionary ethics have been much criticized, both by disbelievers from without and by adherents to one particular position from within. At times, the intragroup warfare resembles the bickering between extreme Protestant sects. Starting with epistemology, and taking first the analogical approach, there are undoubtedly things to recommend it. The growth of knowledge is, in important respects, evolutionary, even Darwinian evolutionary (Adams, 1979; Richards, 1987). Moreover, in important respects, scientific theories or disciplines are much like biological entities, particularly species (whether these be groups or wholes). As David Hull (1981a, 1982a, 1984b, 1988a), particularly, has stressed, a theory like Darwinism is less a body of accepted doctrine or a group of people all holding exactly the same ideas than a diffuse network of beliefs, some consistent, some not, held in varying degrees by varying people — all very much like the organisms in a species (see also, Griesemer, 1984). The big disanalogy, one to which everyone is sensitive, is that the "raw stuff" of biological evolution (i.e., mutations) is random, in the sense that it does not occur according to need (Cohen, 1973, 1974; Thagard, 1980; Amundson, 1983; Kary, 1983; Hookway, 1984b). The raw stuff of human knowledge (i.e., new ideas and discoveries) is usually anything but random. It comes as needed, after careful thought and effort. Whether this is too much of a disanalogy is a moot point. Some think it is; others think not. But it is a worry.

One interesting move was made by Thomas Kuhn (in his *Structure of Scientific Revolutions*) when he argued that, like biological evolution, the direction of science is without overall direction. There is no better or worse, truer or falser, higher or lower. However, accepting that biological evolution has no direction, many would argue that the whole point about science is that it is progressive, toward the truth, even if it never gets there. In any case, given Kuhn's theory that scientific revolutions require paradigm changes, which are abrupt switches from one way of seeing things to another, many feel that there is something a bit odd about Kuhn appealing to evolutionary biology in the first place.

The literal approach to evolutionary epistemology, relying on innate dispositions, strikes some as more promising — even if these same thinkers reject the analogical approach. (This stance contrasts with that of some antireductionists, who like the analogical approach but dislike the literal approach.) However, the literal approach is really little more than a program, at present. It is hardly a fully articulated theory. Much work needs doing to articulate the required dispositions, not to mention the effort required to find empirical backing for the whole approach. Interestingly, following a seminal discussion by the Nobel Prize-winning ethologist Konrad Lorenz (1941), this approach has been embraced with some enthusiasm by a number of scholars (biologists and philosophers) in Austria and Germany (Riedl, 1980; Wuketits, 1978, 1983). Somewhat naturally, they are inclined to link their theorizing with the tradition of the greatest of German philosophers, Immanuel Kant (1929), who likewise argued that the mind is an essential structuring element in human knowledge.

But, although this is an attractive move, others doubt that any Darwinian could truly be said to be a follower of Kant. He sought a necessity to knowledge which transcends the contingencies of the individual human, whereas the Darwinian, seeing knowledge as no more than an adaptation, ultimately has to accept that had our evolution been different, our beliefs would have been different. To get from *A* to *B*, humans walk, horses run, birds fly, monkeys swing through trees, snakes slither, and fish swim. Who is to say that a belief in Euclidean geometry is the only handle to perceptual space? If one is to seek support from the past, these evolutionary epistemologists think that the eighteenth-century Scottish philosopher David Hume (1978) would have been a more promising mentor. Not only was Darwin influenced by his ideas, but Hume argued for a psychological view of knowledge that seems much in tune with contemporary Darwinism. (For other objections to the Kant-Lorenz type of position, see Lewontin, 1982, and Putnam 1982.)

One thing which this debate between Kantians and Humeans does bring to the fore is a problem which has much engaged the general philosophy of science community recently. Is there a real world "out there," independent of our experience? (Does the tree in the forest make a sound if there is no one around to hear it fall?) Kantians incline to think that there is such a reality (the *Ding an Sich*); although, recognizing that we can never experience it, the evolutionary epistemologists tend to follow Popper in speaking of a "hypothetical" reality (Radnitzky and Bartley, 1987). Humeans deny that there is such a reality, arguing that what we see is what we get — bolstered by our innate dispositions to create a reality from our perceptions (Clark, 1986; Ruse, 1986c). Clearly the Humeans avoid the paradoxes of the Kantians, believing in that which is, in principle, beyond experience. Critics, however, think their position no less paradoxical, for apparently it denies that the evolutionary process by which we came into being exists outside our experiencing or thinking about it in some way. (See also Fetzer, 1985, for further discussion.)

Turning now to ethics, for all its distinguished advocates, we find that the position which would base moral justification on the evolutionary process continues to find little favor — even with those generally disposed towards a biological approach to morality (although see Richards, 1986a,b). Apart from difficulties with the whole notion of biological progress (Wachbroit, 1986) — the very essence of Darwinism is that evolution is going nowhere, and rather slowly at that (although see Ayala, 1974c; Nitecki, 1988) — there is a consensus that this kind of endeavor is simply fallacious. One is trying to go from the way things *are* — humans evolved — to the way things *ought* to be — one should care about one's fellows. One is going from statements about matters of fact to statements about directives — from descriptions to prescriptions. (Technically, this is known as a violation of Hume's law, since it was David Hume who first identified and criticized this kind of move [Mackie, 1980].) (See Voorzanger, 1987a; Trigg, 1986; Thomas, 1986; Hughes, 1986; Gewirth, 1986: Cela-Conde, 1986; Ayala, 1987b,c; Alexander, 1980; Campbell, 1979; Caplan, 1978c.)

Probably, matters are a little more complex than this. As with the analytic-synthetic distinction, in recent years the fact-value distinction has come under strong attack (Hudson, 1970). However, no one has yet shown how the collapse of this distinction might save the traditional approach to evolutionary ethics. What of the other approach, which finds morality embedded within adaptive innate dispositions? Even though some find this approach promising, as before, one has something which is more suggestive than fully articulated. One criticism which is often leveled is that no approach allied with sociobiology (which this surely is) could throw light on true morality (Flanagan, 1981; Midgley, 1979; Kitcher, 1985a). Since the sociobiologist starts with the premise that selection acts ultimately for the benefit of the individual (that the genes are "selfish," to recall Dawkins's [1976] notorious but memorable metaphor), this means that morality reduces to enlightened self-interest. And this, of course, is not true morality.

There is no doubt that sociobiologists and perhaps evolutionary ethicists sometimes do get seduced this way. Certainly, such ethicists tend to doubt that we could be unlimited altruists. Yet they would deny that the charge has full force. Causally, the genes may indeed be selfish, in some sense. Yet this is not to say that individual humans are selfish (Ruse, 1976a, 1986b,c). In fact, as hinted already, biology often works better if we do not recognize the full story (Trivers, 1976). The essence of this approach to evolutionary ethics is that we will cooperate only if the morality is genuine, however caused. If we were forever scheming about manipulating people to our own ends, under the guise of morality, we would be far less efficient cooperators than if we were (as indeed we are) genuinely moral. (Although see Moore, 1987, for a well-argued denial that one does ever get genuine altruism.)

So much, then, for the traditional problems of philosophy. Whether evolutionary ethics, or evolutionary epistemology, for that matter, will work is still to be decided. My hunch is that biology has much to offer philosophy. But there is a long way to go.

CHAPTER 9

OTHER
TOPICS

A s I draw toward the end of this review, I have a growing
sense of worry — a worry which reflects well on the subject
of the philosophy of biology and rather less so on this review.
The subject is growing rapidly in size and sophistication and is
already getting beyond the bounds of one review — more
accurately, getting beyond the bounds of one certain reviewer.
In at least two major respects I fear I have done the subject —
and its practitioners — less than justice. Let me try now, very
briefly, to make amends.

First, obviously, rather than "the philosophy of biology
today," my review should more properly be called the "philosophy
of certain parts of evolutionary biology today," given the distorted

selection of subjects that I have covered. To a certain extent, however, the distortion is not entirely mine — some topics do invite philosophical inquiry more than others, and the simple fact of the matter is that philosophers and philosophically minded biologists have written more on these topics than on others. Evolutionary studies, very broadly construed (going from molecular genetics all the way across to paleontology), have excited people, for all sorts of obvious reasons. Conversely, some subjects, popular in the past, seem to get little philosophical notice today — extraterrestrials for instance (although, see Regis, 1985).

But, beyond apology, let me note that other subjects unmentioned or little mentioned by me are getting — or starting to get — philosophical attention. One such area is ecology, where there is growing realization that there is a host of interesting questions about model building, testing, the relationship to the rest of the evolutionary theory, and more (Saarien, 1982; Sloep, 1986; Haila, 1988; Griesemer, 1988). One particular controversy which has attracted some philosophical attention is about the use of null hypotheses and their relation to proper scientific methodology (which should, in the opinion of some, and should not, in the opinion of others, be Popperian) (Simberloff, 1982; Strong, 1982; Beatty, 1987b).

Another area which has been neglected is embryology and development. Most philosophers feel guilty about the topics, promising to investigate but rarely doing so. Perhaps — given the importance of embryology to the vitalists, who saw in the unfolding organism clear evidence of nonmaterial forces — history is the main reason that today's philosophers have tended to steer clear of the subject. However, somewhat paradoxically, given that Darwin was more proud of his embryological explanations than of any other, with the philosophical enthusiasm for non-Darwinian attitudes toward evolution there is now more prospect of attention being paid to embryology. Gould's thinking, for instance, gives a central role to constraints on development, and, as we have seen, this whole approach has generated considerable excitement in recent years. We can expect to see strong moves, by philosophers, in this direction in the future. (See, for instance, Maynard Smith et al., 1985; Ulanowicz, 1986; Wimsatt, 1985b,

1986; Wimsatt and Schank, 1988; Glassmann and Wimsatt, 1984; Schank and Wimsatt, 1987; Burian, 1986b; Kauffman, 1986; Patterson, 1983; and contributions to Bechtel, 1986d.)

Yet a third area which should be mentioned is brain science, in a rather broad sort of sense. Questions to do with the mind and with its relationship to the body are among the oldest in philosophy, and have generated a truly vast literature. In this century, with the rise of the social sciences, we have had considerable philosophical interest in the implications of these subjects — particularly the implications of psychology — for our understanding of the old questions. And recently, in particular, there has been a growing sentiment that biological studies of anatomy and related subjects have an important role to play in such understanding. This is a sentiment which has been complemented by a corresponding feeling about the importance of work in artificial intelligence. It would be wrong to say that traditional workers in the field have uniformly welcomed, with open arms, the influx of biological ideas. Indeed, there are those who feel that the purity of philosophy can only be contaminated by empirical studies. But, like King Canute and the waves, one senses that the tide is coming in, no matter what. Hence, here is an area in which philosophers of biology have the potential to make a considerable impact on other areas in the philosophical spectrum. One hopes that the potential will be fully actualized. (See Churchland, Paul M., 1984, 1986; Churchland, Patricia S., 1986; Morowitz, 1987; Hookway, 1984a; Tennant, 1984; Stillings et al., 1987; Young, 1987; Richardson, 1986b; Anderson, 1986; Rumbaugh, 1986; Savage- Rumbaugh and Hopkins, 1986; Darden, 1983; Dennett, 1987.)

Changing direction somewhat, let me pick up on the fact that in recent years the philosophy of biology is not the only subject that has developed, both in extent and in sophistication. As part of a general maturation of the subject of the history of science in general, the history of biology in particular has grown and developed by leaps and strides. Inevitably, there has been interaction between history and philosophy, with the latter seeing itself as having two tasks. First, there has been the problem of seeing how philosophical ideas functioned in the history of biology.

How, for instance, did the biologists of the day think of philosophy, and did it influence their work? I have mentioned already the gains in our knowledge of Darwin's work which have come through our appreciation of his philosophical sensitivities (Hull, 1973a; Hodge, 1977; Ruse, 1975b; Oldroyd, 1986). Second, what can the history of biology tell us about the nature of scientific change in general? Again, I have touched already on the evolutionary epistemologists who think that biology feeds right into an understanding of scientific change, and some have in fact used biology as a source of case studies (Adams, 1979; Richards, 1977; Bechtel, 1984; Hull, 1988a). Yet this is but one attempt within a wider inquiry. Whether or not one thinks that biology tells about the mechanisms of change, the subject is there as a datum to be explained.

I should add, however, that this area has not progressed quite as rapidly or as fully as one might have hoped. Historians have rightly criticized philosophers for their hurried and naive readings of history (Rudwick, 1986). Philosophers, perhaps also rightly, have criticized historians for their inability to grasp philosophical subtleties and for their too-ready acceptance of relativism. But, now perhaps that both sides have achieved a certain level of security, they can settle into profitable interaction — and there are indeed signs that this is happening. (See, for instance, Kohn, 1985.)

Finally, talking of topics within the philosophy of biology, I must note that — evolutionary ethics apart — there has been growing interest in the connections between biology and values. On the one hand, there are matters to do with values built into biology itself. I have touched on some of these questions in discussing the sociobiology controversy — the extent to which the subject is inherently sexist and racist and so forth. These are part of a broader spectrum of issues about value presuppositions and the like. For instance, interest now is rising in the question of the extent that biology functions as a kind of quasi religion for some scientists, given the way in which they read out meaning and progress and so forth from their subject (Midgley, 1985; Ruse, 1989).

On the other hand, there are external questions of value

— about biology and its status as a force for good and ill. In the last decade the big controversy was about the dangers of recombinant DNA research, about whether or not it should be banned or permitted, even encouraged. Obviously this debate raised issues on which philosophers were qualified to judge — and they did (Richards, 1978; Grobstein, 1979; Stich, 1978). Related questions are about conservation and pollution and so forth, on which there is a veritable cottage industry (Rolston, 1986; Taylor, 1986). And if one's gaze is raised to cover the biomedical sciences, then one is looking at a simply huge enterprise. The same is true if one turns and looks more at questions to do with animal rights and the like (Regan, 1983; Singer, 1975).

CHAPTER 10

OTHER
LANDS

So much for undiscussed topics. The second respect in which I have done the philosophy of biology less than full justice is in the way I have gazed exclusively at English-speaking authors, particularly those from North America. As it happens, in the rest of the English-speaking world (referring especially to Britain, Australasia, and South Africa) there is not much original work being produced on the philosophy of biology — by philosophers, at least. England has long had a tradition of philosophically minded biologists — John Maynard Smith and Richard Dawkins are but two of today's distinguished representatives (to which should be added the articulate Marxist biologist, Steven Rose; see especially, Rose and Rose, 1976 and 1982). But, other than Popper, with his interest in evolutionary epistemology, and one or two others whose interests tend towards the ethical or religious, there is little high-quality work. Tennant (1983a,b, 1986, 1987) is an exception.

Crossing to the continent, however, things are different. If, for the moment, I may change stools from that of disinterested reviewer to that of editor of *Biology and Philosophy,* I must remark that the biggest surprise in the first two years of its existence has been the very large number of submissions from Europe, both West and East. Beginning at the top, Scandinavia has strong links with Anglo-Saxon philosophy (and almost everything is written in English). There is much interest in ecology in Finland, and it is from there that the already-referred-to small but growing literature on the philosophy of ecology is primarily emanating (Saarinen, 1982; Haila, 1986; Haila and Jarrinen, 1982). Norway has first-class philosophers and biologists. Recently they have been collaborating in the fight against Creationism, which has been imported from the United States into that country (Stenseth et al., 1988). In Denmark also there are people active, although, as one might expect from the country of Kierkegaard, much of the interest in biology veers more toward its relation with religion rather than toward straight philosophy (Mortensen and Sorensen, 1987).

Holland has a small subdiscipline which seems virtually unique to that country. So-called theoretical biology runs the gamut from hard-line mathematical modeling to serious study of the philosophical foundations of biology. To date, the subject's practitioners have perhaps been more successful at the mathematical end of the spectrum, thanks to various techniques for capturing the essence of development devised by Aristid Lindenmayer (1973, 1975). Now, however, under the leadership of Wim van der Steen (1983a,b, 1986a,b), a new generation of philosophically trained biologists is producing work on ethics, ecology, theory structure, and more (Sloep, 1986; Voorzanger, 1987a,b).

France has little interest in the subject, but Spain has a number of scholars working on ethics and epistemology from a biological perspective. (See especially Cela-Conde, 1985.) The influence of the Iberian peninsula is felt in South American also, where there is in addition much effort being put into the translation of books authored by English-speaking philosophers. And Italy, especially under the leadership of Vittorio Somenzi, has now contributed to the subject, particularly to such issues as modeling in biology (Cordeschi, 1985, 1986; Stanzione, 1987).

West Germany and Austria have been mentioned already as places sympathetic to evolutionary epistemology (Vollmer, 1975), and in both Germanies there is keen interest in the philosophical foundations of evolutionary theory, particularly that approach which takes form as primary over function (Grene, 1983). Very much worthy of mention is the fact that every three years in the German Democratic Republic there is a week-long meeting bringing together philosophers and biologists (the "Kullungsborn conference"); recent topics discussed have included the philosophical significance of Darwinism and human sociobiology (Geissler and Scheler, 1983). In Poland there is much interest in philosophical problems of paleontology, under the guidance of Adam Urbanek (see also Hoffman, 1983), and Czechoslovakia (under the guidance of Vladimir Novak) has hosted a series of conferences on development and evolutionary thought, paying special attention to philosophical questions (Mlikovsky and Novak, 1985).

Finally, let me mention the USSR. As is well known, biology there has had a somewhat checkered career. Thanks to the praise (albeit not always unqualified) of Marx and Lenin, Darwin has always had an honored place in Soviet thought. However, during the Stalin and Kruschev regimes, science took a back place to politics. But since the early 1960s, biology has moved back on course, and the philosophy of biology has thrived (Graham, 1972, 1981, 1987). This interest has come about particularly through the influence of Ivan Frolov, who was strong in the attack on Lysenko. As the editor for several years of the leading Soviet philosophy journal, *Voprosy Filosofii*, Frolov has encouraged a group of younger workers. In recent years, there has been much interest in human sociobiology (Satdinova, 1982; Frolov, 1986). Naturally, as Marxist-Leninists, Soviet thinkers tend to be critical — but, as one who has been at the receiving end of criticism, I must say (and I do not now intend to be insufferably condescending) that the level of understanding and awareness is remarkably high. In contrast to certain American critics, one senses a genuine attempt to grasp the other's viewpoint. (See also Frolov, 1977.)

Other countries make some contributions, from Israel through Japan (Kawata, 1987) to China (Boshu, 1987). I trust that this brief survey has given some idea of the work which is going on across the globe and makes some amends for the inadequacies of my review.

Conclusion

I have worked as a philosopher of biology for twenty years. To use a biological metaphor, I have seen it grow from little more than an embryo to a gangling adolescent. I hope I can stay to see the mature beast, and I hope also that I have shown why I feel this way. Despite disagreements — because of disagreements — the philosophy of biology today is indeed a hot topic, worthy of attention.

Bibliography

Achinstein, P. (1977) Function statements. *Philosophy of Science*, 44: 341-67.

Adams, M.B. (1979) From "Gene fund" to "Gene pool": On the evolution of evolutionary language. In W. Coleman and C. Limoges, eds., *Studies in the History of Biology*, 3: 241-85.

Alexander, R.D. (1976) *Evolution, Human Behavior, and Determinism.* In F. Suppe and P. Asquith, eds., *PSA 1976* (East Lansing, Mich.: Philosophy of Science Association), 2: 3-21.

—— (1979) *Darwinism and Human Affairs* (Seattle: University of Washington Press).

—— (1980) Evolution, social behavior and ethics. In H.T. Engelhardt, ed., *Knowing and Valuing: The Search for Common Roots* (Hastings on Hudson, N.Y.: Hastings Center), 124-55.

—— (1987) *The Biology of Moral Systems* (New York: Aldine de Gruyter).

Allen, E., et al.(1977) Sociobiology: A new biological determinism. In Sociobiology Study Group of Boston, ed., *Biology as a Social Weapon* (Minneapolis: Burgess).

Amundson, R. (1983) The epistemological status of naturalized epistemology. *Inquiry*, 26: 333-44.

Anderson, R.A. (1986) Cognitive explanations and cognitive ethology. In W. Bechtel, ed., *Integrating Scientific Disciplines* (Dordrecht, Holland: Nijhoff), 309-22.

Arnold, A.J., and K. Fristrup (1984) The theory of evolution by natural selection: A hierarchical expansion. In R.N. Brandon and R.M. Burian, eds., *Genes, Organisms and Populations* (Cambridge, Mass.: The MIT Press), 292-320.

Ayala, F.J. (1968) Biology as an autonomous science. *American Scientist*, 56: 207-21.

—— (1970a) Teleological explanations in evolutionary biology. *Philosophy of Science*, 37: 1-15.

—— (1970b) Comments on methodology in the physical, biological and social sciences. In E.O. Attinger, ed., *Global Systems Dynamics* (New York: Karger), 28-33.

—— (1971) A biologist's view of nature. In G.C. Stone, ed., *A New Ethic for a New Earth* (New York: Friendship Press), 24-39.

—— (1972a) The evolutionary thought of Teilhard de Chardin. In A.D. Breck and W. Yourgrau, eds., *Biology, History, and Natural Philosophy* (New York: Plenum), 207-16.

—— (1972b) Darwinian versus non-Darwinian evolution in natural populations of *Drosophila. Proceedings of the Sixth Berkeley Symposium on Mathematics, Statistics and Probability*, 5: 211-36.

—— (1973) Man in evolution. In J.N. Deely and R.J. Nogar, eds., *The Problem of Evolution* (New York: Appleton-Century-Crofts), 237-50.

—— (1974a) Genetic differentiation within and between species of the Drosophila willistoni group. *Proceedings of the National Academy of Science*, 71: 999-1003.

—— (1974b) Introduction. In F.J. Ayala and Th. Dobzhansky, eds., *Studies in the Philosophy of Biology* (London: Macmillan), vii-xvi.

—— (1974c) The concept of biological progress. In F.J. Ayala and Th. Dobzhansky, eds., *Studies in the Philosophy of Biology* (London: Macmillan), 339-54.

—— (1974d) Biological evolution: natural selection or random walk? *American Scientist*, 62: 692-701.

—— (1975a) Scientific hypothesis, natural selection and the neutrality theory of protein evolution. In F.M. Salzano, ed., *The Role of Natural Selection in Human Evolution* (Elsevier: Amsterdam), 19-42.

—— (1975b) Genetic differentiation during the speciation process. *Evolutionary Biology*, 8: 1-78.

—— (1982) The evolutionary concept of progress. In G.A. Almond, M.Chodorow and R.H. Pearce, eds., *Progress and Its Discontents* (Berkeley: University of California Press), 106-24.

—— (1983a) Microevolution and macroevolution. In D.S. Bendall, ed., *Evolution from Molecules to Men* (Cambridge: Cambridge University Press), 387-402.

—— (1983b) Biology and physics: reflections on reductionism. In A. van der Merwe, ed., *Old and New Questions in Physics, Cosmology, Philosophy, and Theoretical Biology: Essays in Honor of Wolfgang Yourgrau* (New York: Plenum), 525-34.

—— (1983c) Beyond Darwinism? The challenge of macroevolution to the synthetic theory of evolution. In P. Asquith and T. Nickles, eds., *PSA 1982* (East Lansing, Mich.: Philosophy of Science Association), 2: 275-91.

—— (1985a) Reduction in biology: A recent challenge. In D.J. Depew and B.H. Weber, eds., *Evolution at a Crossroads* (Cambridge, Mass.: The MIT Press), 65-80.

—— (1985b) The theory of evolution: Recent successes and challenges. In E. McMullin, ed., *Evolution and Creation* (Notre Dame, Ind.: University of Notre Dame Press), 59-90.

—— (1986) Whither mankind? The choice between a genetic twilight and a moral twilight. *American Zoologist*, 26: 895-905.

—— (1987a) Biological reductionism: the problems and some answers. In F.E. Yates, ed., *Self-Organizing Systems: The Emergence of Order* (New York: Plenum), 315-24.

—— (1987b) Sociobiology and ethics. *History and Philosophy of the Life Sciences*, 9: 321-31.

—— (1987c) The biological roots of morality. *Biology and Philosophy*, 2: 235-52.

Ayala, F., and Th. Dobzhansky, eds. (1974) *Studies in the Philosophy of Biology* (Berkeley: University of California Press).

Ayala, F.J., M.L. Tracey, L.G. Barr, J.F. McDonald, and S. Perez-Salas (1974) Genetic variation in natural populations of five Drosophila species and the hypothesis of the selective neutrality of protein polymorphisms. *Genetics*, 77: 343-84.

Ayala, F.J., and J.W. Valentine (1979) *Evolving* (Menlo Park, Calif.: Benjamin/Cummings).

Ayers, M.R. (1981) Locke versus Aristotle on natural kinds. *Journal of Philosophy*, 78: 247-72.

Balzer, W., and C.M. Dawe (1986a) Structure and comparison of genetic theories: (1) Classical genetics. *British Journal for the Philosophy of Science*, 37: 55-69.

—— (1986b) Structure and comparison of genetic theories: (2) The reduction of character-factor genetics to molecular genetics. *British Journal for the Philosophy of Science*, 37: 171-91.

Barash, D. (1982) *Sociobiology and Behavior*, 2nd ed., (New York: Elsevier).

Barlow, G.W., and J. Silverberg, eds., (1980) *Sociobiology: Beyond Nature/Nurture?* (Boulder, Col.: Westview).

Baublys, K.K. (1975) Discussion: Comments on some recent analyses of functional statements in biology. *Philosophy of Science* , 42: 469-86.

Beatty, J. (1978) *Evolution and the Semantic View of Theories*. Ph.D. thesis. Indiana University.

—— (1980) Optimal-design models and the strategy of model building in evolutionary biology. *Philosophy of Science*, 47: 532-61.

—— (1981) What's wrong with the received view of evolutionary theory? In P.D. Asquith and R.M. Giere, eds., *PSA 1980* (East Lansing, Mich.: Philosophy of Science Association), 2: 397-426.

—— (1982a) Classes and cladists. *Systematic Zoology*, 31: 25-34.

—— (1982b) What's in a word? Coming to terms in the Darwinian Revolution. *Journal of the History of Biology*, 15: 215-39.

—— (1983) The insights and oversights of molecular genetics: The place of the evolutionary perspective. In P. Asquith and T. Nickles, eds., *PSA 1982* (East Lansing, Mich.: Philosophy of Science Association), 2: 341-55.

—— (1984) Chance and natural selection. *Philosophy of Science,* 51 183-211.

—— (1985a) Speaking of species: Darwin's strategy. In D. Kohn, ed., *The Darwinian Heritage* (Princeton N.J.: Princeton University Press), 265-81.

—— (1985b) Pluralism and panselectionism. In P. Asquith and P. Kitcher , eds., *PSA 1984* (East Lansing, Mich.: Philosophy of Science Association), 2, 113-28.

—— (1986) The synthesis and the synthetic theory. In W. Bechtel, ed., *Integrating Scientific Disciplines* (Dordrecht, Holland: Nijhoff), 125-36.

—— (1987a) Dobzhansky and drift: Facts, values, and chance in evolutionary biology. In L. Kruger, ed., *The Probabilistic Revolution* (Cambridge, Mass.: The MIT Press).

—— (1987b) Natural selection and the null hypothesis. In J. Dupre, ed., *The Latest on the Best* (Cambridge, Mass.: The MIT Press), 53-75.

—— (1987c) On behalf of the semantic view. *Biology and Philosophy,* 2: 17-22.

—— (1987d) Weighing the risks: Stalemate in the Classical/ Balance controversy. *Journal of the History of Biology,* 20: 289-319.

Bechtel, W. (1982a) Two common errors in explaining biological and psychological phenomona. *Philosophy of Science,* 49: 549-74.

—— (1982b) Reconceptualizations and interfield connections: The discovery of the link between vitamins and coenzymes. *Philosophy of Science,* 51: 265-92.

—— (1983) Forms of organization and the incompletability of science. In N. Rescher, ed., *The Limits of Lawfulness* (Landham, Md.: University Press of America), 79-92.

—— (1984) The evolution of our understanding of the cell: A study in the dynamics of scientific progress. *Studies in History and Philosophy of Science,* 15: 309-56.

—— (1986a) Building interlevel theories: The discovery of the Embden-Meyerhof pathway and the phosphate cycle. In P. Weingartner and G. Dorn, eds., *Foundations of Biology* (Vienna: Holder-Pichler-Tempsky), 65-96.

—— (1986b) Biochemistry: A disciplinary endeavor that discovered a distinctive domain. In W. Bechtel, ed., *Integrating Scientific Disciplines* (Dordrecht, Holland: Nijhoff), 77-100.

—— (1986c) Teleological functional analysis and the hierarchical organization of nature. In N. Rescher, ed., *Current Issues in Teleology* (Landham, Md.: University Press of America), 26-48.

—— , ed. (1986d) *Integrating Scientific Disciplines* (Dordrecht, Holland: Nijhoff).

Beckner, M. (1959) *The Biological Way of Thought* (New York: Columbia University Press).

—— (1969) Function and teleology. *Journal of the History of Biology*, 2: 151-64.

Bergson, H. (1913) *Creative Evolution* (London: Macmillan).

Berry, R.J. (1986) What to believe about miracles. *Nature*, 322: 321-22.

Bigelow, J., and R. Pargetter (1987) Functions. *Journal of Philosophy*, 84: 181-96.

Birke, L. (1986) *Women, Feminism and Biology* (Brighton, Gt. Britain: Wheatsheaf).

Bleier, R. (1984) *Science and Gender: A Critique of Biology and Its Theories on Women* (New York: Pergamon Press).

Bock, W.J. (1986) Species concepts, speciation, and macroevolution. In K. Iwatsuki, P.H. Raven and W.J. Bock, eds., *Modern Aspects of Species* (Tokyo: University of Tokyo Press), 31-57.

Bonner, J.T. (1980) *The Evolution of Culture in Animals* (Princeton, N.J.: Princeton University Press).

Borgerhoff Mulder, M. (1987) Adaptation and evolutionary approaches to anthropology. *Man*, 22: 25-41.

Boshu, Z. (1987) Marxism and human sociobiology: A comparative study from the perspective of modern socialist economic reforms. *Biology and Philosophy*, 2: 463-74.

Bowler, P. (1986) *Theories of Human Evolution* (Baltimore, Md.: John Hopkins University Press).

Boyd, R., and P.J. Richerson (1985) *Culture and the Evolutionary Process* (Chicago: University of Chicago Press).

Bradie, M. (1986) Assessing evolutionary epistemology. *Biology and Philosophy*, 1: 401-59.

Brady, R.H. (1982) Theoretical issues and pattern cladistics. *Systematic Zoology*, 31: 286-91.

Brandon, R.N. (1978a) Adaptation and evolutionary theory. *Studies in History and Philosophy of Science*, 9: 181-206.

—— (1978b) Evolution. *Philosophy of Science*, 45: 96-109.

—— (1981a) A structural description of evolutionary theory. In P. Asquith and R. Giere, eds., *PSA 1980* (East Lansing, Mich.: Philosophy of Science Association), 2: 427-39.

—— (1981b) Biological teleology: Questions and explanations. *Studies in History and Philosophy of Science*, 12: 91-105.

—— (1983) Levels of selection. In P. Asquith and T. Nickles, eds., *PSA 1982* (East Lansing, Mich.: Philosophy of Science Association), 2: 315-23.

—— (1985a) Adaptation explanations: Are adaptations for the good of replicators or interactors? In D.J. Depew and B.H. Weber, eds., *Evolution at a Crossroads* (Cambridge, Mass.: The MIT Press), 81-96.

—— (1985b) Grene on mechanism and reductionism: More than just a side issue. In P. Asquith and P. Kitcher, eds., *PSA 1984* (East Lansing, Mich.: Philosophy of Science Association), 2: 345-53.

—— (1985c) Holism in philosophy of biology. In D. Stalker and C. Glymour, eds., *Holistic Medicine: A Critical Examination* (Buffalo, N.Y.: Prometheus), 127-36.

—— (1985d) Plasticity, cultural transmission and human sociobiology. In J. Fetzer, ed., *Sociobiology and Epistemology* (Dordrecht, Holland: Reidel), 57-73.

Brandon R. N., and J. Beatty (1984) Discussion: The propensity interpretation of 'fitness' — no interpretation is no substitute. *Philosophy of Science*, 51: 342-47.

Brandon, R.N., and R.M. Burian, eds. (1984). *Genes, Organisms, Populations: Controversies Over the Units of Selection* (Cambridge, Mass.: The MIT Press).

Brandon, R.N., and N. Hornstein (1986) From icons to symbols:

Some speculations on the origins of language. *Biology and Philosophy*, 1: 169-90.

Brewster, D. (1854) *More Worlds than One: The Creed of the Philosopher and the Hope of the Christian* (London, Murray).

Brooks, D.R. (1983) What's going on in evolution? A brief guide to some new ideas in evolutionary theory. *Canadian Journal of Zoology*, 61: 2637-45.

Brooks, D.R., and R.T. O'Grady (1986) Nonequilibrium thermo-dynamics and different axioms of evolution. *Acta Biotheoretica*, 35: 77-106.

Brooks, D.R., and E.O. Wiley (1984) Evolution as an entropic phenomenon. In J. Pollard, ed., *Evolutionary Theory: Paths into the Future* (Chichester, Gt. Britain: J. Wiley and Sons), 141-72.

—— (1986) *Evolution as Entropy: Toward a Unified Theory of Biology* (Chicago: University of Chicago Press).

Buck, R., and D.L. Hull (1966) The logical structure of the Linnean hierarchy. *Systematic Zoology*, 15: 97-111.

Burian, R.M. (1981) Human sociobiology and genetic determinism. *Philosophical Forum*, 13 (2-3): 43-66.

—— (1983) Adaptation. In M. Grene, ed., *Dimensions of Darwinism* (Cambridge: Cambridge University Press), 287-314.

—— (1985) On conceptual change in biology: The case of the gene. In D.J. Depew and B.H. Weber, eds., *Evolution at a Crossroads* (Cambridge, Mass.: The MIT Press), 21-42.

—— (1986a) Why the panda provides no comfort to the Creationist. *Philosophica*, 37: 11-26.

—— (1986b) On integrating the study of evolution and development. In W. Bechtel, ed., *Integrating Scientific Disciplines* (Dordrecht, Holland: Nijhoff), 209-28.

Byerly, H. (1986) Fitness as function. In P. Asquith and P. Kitcher, eds., *PSA 1986* (East Lansing, Mich.: Philosophy of Science Association), 1: 494-501.

Cain, A.J. (1979) Introduction to general discussion [of Gould and Lewontin: Spandrels of San Marco]. *Proceedings of the Royal Society of London*, Series B, 205: 599-604.

Callebaut, W., ed. (1986) *Current Issues in the Philosophy of Biology.* In *Philosophica*, 37: 1-162.

Campbell, D.T. (1973) Ostensive instances and entitativity in language learning. In W. Gray and N.D. Rizzo, ed., *Unity Through Diversity* (New York: Gordon and Breach), 2: 1043-57.

—— (1974a) Unjustified variation and selective retention in scientific discovery. In F. Ayala and Th. Dobzhansky, eds., *Studies in the Philosophy of Biology* (London: Macmillan), 139-61.

—— (1974b) 'Downward causation' in hierarchically organized biological systems. In F. Ayala and Th. Dobzhansky, eds., *Studies in the Philosophy of Biology* (London: Macmilan), 179-86.

—— (1974c) Evolutionary epistemology. In P.A. Schilpp, ed., *The Philosophy of Karl Popper* (LaSalle, Ill.: Open Court Publishing), 1: 413-63.

—— (1975) On the conflicts between biological and social evolution and between psychology and moral tradition. *American Psychologist*, 30: 1103-26.

—— (1977) Discussion: Comment on the natural selection model of conceptual evolution. *Philosophy of Science*, 44: 502-7.

—— (1979) Comments on the sociobiology of ethics and moralizing. *Behavioural Science*, 24: 37-45.

—— (1983) The two distinct routes beyond kin selection to ultrasociality: Implications for the humanities and social sciences. In D.C. Bridgeman, ed., *The Nature of Prosocial Development: Theories and Strategies* (New York: Academic Press), 11-41.

—— (1986) Rationality and utility from the standpoint of evolutionary biology. *Journal of Business*, 59 (4): 355-64.

—— (1987a) Neurological embodiments of belief and the gaps in the fit of phenomena to noumena. In A. Shimony and D. Nails, eds., *Naturalistic Epistemology: A Symposium of Two Decades* (Dordrecht, Holland: Reidel), 165-92.

—— (1987b) Selection theory and the sociology of scientific validity. In W.G. Callebaut and R. Pinxten, eds., *Evolutionary*

238509

Epistemology: A Multiparadigm Program (Dordrecht, Holland: Reidel), 139-58.

—— (1988a) A general 'selection theory' as implemented in biological evolution and in social belief — transmission-with-modification in science. *Biology and Philosophy*, 3: 171-7.

—— (1988b) Levels of organization, downward causation, and the selection-theory approach to evolutionary epistemology. In E. Tobach and G. Greenberg, eds., *Scientific Methodology in the Study of Mind: Evolutionary Epistemology* (Hillsdale, N.J.: Lawrence Erlbaum).

Campbell, D.T., C.M. Heyes, and W.G. Callebaut (1986) Evolutionary epistemology bibliography. In W.G. Callebaut and R. Pinxten, eds., *Evolutionary Epistemology: A Multiparadigm Program* (Dordrecht, Holland: Reidel).

Campbell, J.H. (1985). An organizational interpretation of evolution. In D.J. Depew and B.H. Weber, eds., *Evolution at a Crossroads* (Cambridge, Mass.: The MIT Press), 133-65.

Caplan, A. (1977) Tautology, circularity, and biological theory. *American Naturalist*, 111: 390-93.

—— , ed. (1978a) *The Sociobiology Debate* (New York: Harper and Row).

—— (1978b) Testability, disreputability, and the structure of the modern synthetic theory of evolution. *Erkenntnis*, 13: 261-78.

—— (1978c) In what ways are recent developments in biology and sociobiology relevant to ethics? *Perspectives in Biology and Medicine*, 21: 536-50.

—— (1979) Darwinism and deductivist models of theory structure. *Studies in History and Philosophy of Science*, 10: 341-53.

—— (1980) Have species become declasse? In P. Asquith and R. Giere, eds., *PSA 1980* (East Lansing, Mich.: Philosophy of Science Association), 1: 71-82.

—— (1981a) Back to class: A note on the ontology of species. *Philosophy of Science*, 48: 130-40.

—— (1981b) Say it just ain't so: Adaptational stories and sociobiological explanations of social behavior. *Philosophical Forum*, 13 (2-3): 144-60.

—— (1981c) Babies, bathwater, and derivational reduction. In P. Asquith and I. Hacking, eds., *PSA 1978* (East Lansing, Mich.: Philosophy of Science Association), 2: 357-70.

—— (1984) Sociobiology as a strategy in science. *The Monist*, 67: 143-60.

—— (1985) Is Darwinian evolutionary theory scientific? In L. Godfrey, ed., *What Darwin Began* (Boston: Allyn and Bacon), 24-29.

—— (1986) Exemplary reasoning? A comment on theory structure in biomedicine. *The Journal of Medicine and Philosophy*, 11: 93-105.

—— (1987) Why the problem of reductionism in biology won't go away. *Growth*, 51: 23-34.

—— (1988) Rehabilitating reductionism. *American Zoologist*, 28: 193-203.

Caplan, A.L., and W. Bock (1988) Spirit — haunt me no more. *Biology and Philosophy*, 3, forthcoming.

Caro, T.M. and M. Borgerhoff Mulder (1987) The problem of adaptation in the study of human behavior. *Ethology and Sociobiology*, 8: 61-72.

Cela-Conde, C.J. (1985) *De Genes, Dioses y Tiranos* (Madrid: Alianza). Translated into English as *On Genes, Gods and Tyrants* (Dordrecht: Reidel, 1987).

—— (1986) The challenge of evolutionary ethics. *Biology and Philosophy*, 1: 293-96.

Churchland, Patricia S. (1986) *Neurophilosophy: Toward a Unified Science of Mind-Brain* (Cambridge, Mass.: The MIT Press).

Churchland, Paul M. (1984) *Matter and Consciousness* (Cambridge, Mass.: The MIT Press).

—— (1986) Cognitive neurobiology: A computational hypothesis for laminar cortex. *Biology and Philosophy*, 1: 25-52.

Clark, A. (1986) Evolutionary epistemology and the scientific method. *Philosophica*, 37: 151-62.

Cohen, L.J. (1973) Is the progress of science evolutionary? *British Journal for the Philosophy of Science*, 24: 41-61.

—— (1974) Professor Hull and the evolution of science. *British Journal for the Philosophy of Science*, 25: 334-36.

Collier, J. (1985) Entropy in evolution. *Biology and Philosophy*, 1: 5-24.

Cordeschi, R. (1985) Mechanical models in psychology in the 1950's. *Studies in the History of Psychology and the Social Sciences*, 3.

—— (1986) Kenneth Craik and the "mechanistic tendency of modern psychology." *Rivista di Storia della Scienza*, 3(2): 233-52.

Cracraft, J. (1974) Phylogenetic models and classification. *Systematic Zoology*, 23: 71-90.

—— (1978) Science, philosophy, and systematics. *Systematic Zoology*, 27: 213-15.

—— (1987) Species concepts and the ontology of evolution. *Biology and Philosophy*, 2: 329-46.

Crowe, T.M. (1987) Species as individuals or classes: An "iconoclassificationist's" view. *Biology and Philosophy*, 2: 167.

Cullis, C.A. (1984) Environmentally induced DNA changes. In J. Pollard, ed., *Evolutionary Theory: Paths into the Future*, (Chichester, Gt. Britain: J. Wiley and Sons), 203-16.

Cummins, R. (1975) Functional analysis. *Journal of Philosophy*, 72: 741-65.

Darden, L. (1980) Theory construction in genetics. In T. Nickles, ed., *Scientific Discovery: Case Studies* (Dordrecht, Holland: Reidel), 151-70.

—— (1983) Artificial intelligence and philosophy of science: Reasoning by analogy in theory construction. In T. Nickles and P. Asquith, eds., *PAS 1982* (East Lansing, Mich.: Philosophy of Science Association), 2: 147-61.

—— (1986a) Relations among fields in the evolutionary synthesis. In W. Bechtel, ed., *Integrating Scientific Disciplines* (Dordrecht, Holland: Nijhoff), 113-24.

—— (1986b) Reasoning in theory construction: Analogies, interfield connections and levels of organization. In P. Weingartner and G. Dorn, eds., *Foundations of Biology* (Vienna: Holder-Pichler-Tempsky), 99-107.

—— (1987) Viewing history of science as compiled hindsight. *AI Magazine*, 8 (2): 33-41.

Darden, L., and J.A. Cain (1988) Selection type theories. *Philosophy of Science*, forthcoming.

Darden, L., and N. Maull (1977) Interfield theories. *Philosophy of Science*, 44: 43-64.

Darwin, C. (1859) *On the Origin of Species* (London: Murray).

Dawkins, R. (1976) *The Selfish Gene* (Oxford: Oxford University Press).

—— (1978) Replicator selection and the extended phenotype. *Zeitschrift fur Tier-psychologie*, 47: 61-76.

—— (1982) *The Extended Phenotype: The Gene as the Unit of Selection* (San Francisco: Freeman).

—— (1983) Universal Darwinism. In D.S. Bendall, ed., *Evolution from Molecules to Men* (Cambridge: Cambridge University Press), 403-25.

—— (1986) *The Blind Watchmaker* (London: Longman).

Dennett, D.C. (1983) Intentional systems in cognitive ethology: The "Panglossian paradigm" defended. *Behavorial and Brain Sciences*, 6: 343-90.

—— (1987) *The Intentional Stance* (Cambridge, Mass.: The MIT Press).

Depew, D.J. (1986) Nonequilibrium thermodynamics and evolution. *Philosophica*, 37: 27-58.

Depew, D.J. and B.H. Weber (1985a) Innovation and tradition in evolutionary theory: An interpretive afterword. In D.J. Depew and B.H. Weber, eds., *Evolution at a Crossroads* (Cambridge, Mass.: The MIT Press), 227-60.

—— , eds. (1985b) *Evolution at a Crossroads* (Cambridge, Mass.: The MIT Press).

de Waal, F. (1982) *Chimpanzee Politics* (London: Cape).

Dobzhansky, Th. (1951) *Genetics and the Origin of Species*, 3rd ed. (New York: Columbia University Press).

—— (1962) *Mankind Evolving* (New Haven, Conn.: Yale University Press).

—— (1970) *Genetics of the Evolutionary Process* (New York: Columbia).

—— (1973) Nothing in biology makes sense except in the light of evolution. *American Biology Teacher*, 35: 125-9.

Dobzhansky, Th. and F.J. Ayala (1977) Humankind — A product of evolutionary transcendence. *Special Raymond Dart Lecture* (Johannesburg: Witwatersrand University Press).

Dobzhansky, Th., F.J. Ayala, G.L. Stebbins, and J.W. Valentine (1977) *Evolution* (San Francisco: Freeman).

Doolittle, W.F. (1985) Some broader evolutionary issues which emerge from contemporary molecular biology. In P. Asquith and P. Kitcher, eds., *PSA 1984* (East Lansing, Mich.: Philosophy of Science Association), 2: 129-44.

Dover, G.A. (1982) Molecular drive: A cohesive mode of species formation. *Nature*, 299: 338-47.

Driesch, H. (1905) *Der Vitalismus als Geschichte und als Lehre* (Leipzig: J.A. Barth).

—— (1914) *The History and Theory of Vitalism.* (London: Macmillan).

Dupré, J. (1981) Natural kinds and biological taxa. *Philosophical Review*, 90: 66-90.

—— ,ed. (1987) *The Latest on the Best: Essays on Evolution and Optimality* (Cambridge, Mass.: The MIT Press).

Dyke, C. (1985) Complexity and closure. In D.J. Depew and B.H. Weber, eds., *Evolution at a Crossroads* (Cambridge, Mass.: The MIT Press), 97-132.

—— (1988) *The Evolutionary Dynamics of Complex Systems* (New York: Oxford University Press).

Eldredge, N. (1971) The allopatric model and phylogeny in Paleozoic invertebrates. *Evolution*, 25: 156-67.

—— (1982) Phenomenological levels and evolutionary rates. *Systematic Zoology*, 31: 338-47

—— (1985a) *Time Frames: The Rethinking of Darwinian Evolution and the Theory of Punctuated Equilibria* (New York: Simon and Schuster).

—— (1985b) *Unfinished Synthesis* (New York: Oxford University Press).

—— (1985c) Large-scale biological entities and the evolutionary process. In P. Asquith and P. Kitcher, eds., *PSA 1984* (East Lansing, Mich.: Philosophy of Science Association), 2: 525-42.

Eldredge, N., and J. Cracraft (1980) *Phylogenetic Patterns and the Evolutionary Process* (New York: Columbia University Press).

Eldredge, N., and S.J. Gould (1972) Punctuated equilibria: An alternative to phyletic gradualism. In T.J.M. Schopf, ed., *Models in Paleobiology* (San Francisco: Freeman Cooper), 82-115.

Endler, J.A. (1986) *Natural Selection in the Wild* (Princeton, N.J.: Princeton University Press).

Fales, E. (1982) Natural kinds and freaks of nature. *Philosophy of Science*, 49: 67-90.

Falk, A.E. (1981) Purpose, feedback and evolution. *Philosophy of Science*, 48: 198-217.

Felsenstein, J., and E. Sober (1986) Likelihood and parsimony. *Systematic Zoology*, 35: 617-26.

Fetzer, J. ed. (1985) *Sociobiology and Epistemology* (Dordrecht, Holland: Reidel).

Flanagan, O.J. (1981) Is morality epiphenomenal? The failure of the sociobiological reduction of ethics. *Philosophical Forum*, 13 (2-3): 207-25.

Friedman, R. (1986) Necessitarianism and teleology in Aristotle's biology. *Biology and Philosophy*, 1: 355-66.

Frolov, I. (1977) Foreword to *The Philosophy of Biology* by Michael Ruse (Moscow: Progress). In Russian.

—— (1986) Genes or culture? A Marxist perspective on humankind. *Biology and Philosophy*, 1: 89-108.

Futuyma, D. (1979) *Evolutionary Biology* (Sunderland, Mass.: Sinauer). Second edition, 1986.

—— (1983) *Science on Trial* (New York: Pantheon).

Gaffney, E.S. (1979) An introduction to the logic of phylogeny reconstruction. In J. Cracraft and N. Eldredge, eds., *Phylogenetic Analysis and Paleontology* (New York: Columbia University Press), 79-111.

Gallie, W.B. (1955) Explanations in history and the genetic sciences. *Mind*, 64: 160-80.

Geissler, E., and W. Scheler, eds. (1983) *Darwin Today* (Berlin: Akademie-Verlag).

Gewirth, A. (1986) The problem of specificity in evolutionary ethics. *Biology and Philosophy*, 1: 297-305.

Ghiselin, M. (1966) On psychologism in the logic of taxonomic controversies. *Systematic Zoology*, 15: 207-15.

—— (1969) *The Triumph of the Darwinian Method* (Berkeley: University of California Press).

—— (1974) A radical solution to the species problem. *Systematic Zoology*, 23: 536-44.

—— (1981) Categories, life, and thinking. *Behavorial and Brain Sciences*, 4: 269-313.

—— (1987) Species concepts, individuality and objectivity. *Biology and Philosophy*, 2: 127-45.

Gibbard, A. (1982) Human evolution and the sense of justice. In P. French et al., eds., *Midwest Studies in Philosophy* (Minneapolis: University of Minnesota Press), 7: 31-46.

Giere, R. (1979) *Understanding Scientific Reasoning* (New York: Holt, Rinehart and Winston).

—— (1985) Philosophy of science naturalized. *Philosophy of Science*, 52: 331-56.

Gifford, F. (1986) Sober's use of unanimity in the units of selection problem. In P. Asquith and P. Kitcher, eds., *PSA 1986* (East Lansing, Mich.: Philosophy of Science Association), 1: 473-83.

Gingerich, P.D. (1976) Paleontology and phylogeny: Patterns of evolution at the species level in early Tertiary mammals. *American Journal of Science*, 276: 1-28.

—— (1977) Patterns of evolution in the mammalian fossil record. In A. Hallam, ed., *Patterns of Evolution, As Illustrated by the Fossil Record* (Amsterdam: Elsevier), 469-500.

Giray, E. (1976) An integrated biological approach to the species problem. *British Journal for the Philosophy of Science*, 27: 317-28.

Gish, D.T. (1972) *Evolution: The Fossils Say No!* (San Diego, Calif.: Creation-Life).

Glassmann, R.B., and W.C. Wimsatt (1984) Evolutionary advantages and limitations of early plasticity. In R. Almli and S. Finger, eds., *Early Brain Damage* (New York: Academic Press), 1: 35-58.

Goodall, J. (1986) *The Chimpanzees of Gombe* (Cambridge, Mass.: Harvard University Press).

Goodwin, B.C. (1984) Changing from an evolutionary to a generative paradigm in biology. In J. Pollard, ed., *Evolutionary Theory: Paths into the Future* (Chichester, Gt. Britain: J. Wiley and Sons), 99-120.

Gotthelf, A. and J. Lennox, eds. (1987) *Philosophical Issues in Aristotle's Biology* (Cambridge: University of Cambridge Press).

Goudge, T.A. (1961) *The Ascent of Life* (Toronto: University of Toronto Press).

Gould, S.J. (1971) D'Arcy Thompson and the science of form. *New Literacy History*, 2: 229-58.

—— (1978) Sociobiology: The art of story-telling. *New Scientist*, 80: 530-33.

—— (1979) Episodic change versus gradualist dogma. *Science and Nature*, 2: 5-12.

—— (1980a) Caring groups and selfish genes. In *The Panda's Thumb* (New York: W.W. Norton), 85-92. Reprinted in E. Sober, ed. (1984) *Conceptual Issues in Evolutionary Biology* (Cambridge, Mass.: The MIT Press), 119-24.

—— (1980b) The promise of paleobiology as a nomothetic, evolutionary discipline. *Paleobiology*, 6: 96-118.

—— (1980c) Is a new and general theory of evolution emerging? *Paleobiology*, 6: 119-30.

—— (1981) *The Mismeasure of Man* (New York: W.W. Norton).

—— (1982a) Darwinism and the expansion of evolutionary theory. *Science*, 216: 380-87.

—— (1982b) Punctuated equilibrium — a different way of seeing. In J. Cherfas, ed., *Darwin Up to Date* (London: IPC Magazines), 26-30.

—— (1982c) The meaning of punctuated equilibrium and its role in validating a hierarchical approach to macroevolution. In R. Milkman, ed., *Perspectives On Evolution* (Sunderland, Mass.: Sinauer), 83-104.

—— (1983a) Irrelevance, submission, and partnership: The changing role of palaeontology in Darwin's three centennials, and a modest proposal for macroevolution. In D.S. Bendall, ed., *Evolution from Molecules to Men* (Cambridge Cambridge University Press), 347-66.

—— (1983b) The hardening of the modern synthesis. In M. Grene, ed., *Dimensions of Darwinism* (Cambridge: Cambridge University Press), 71-93.

Gould, S.J., and N. Eldredge (1977) Punctuated equilibria: The tempo and mode of evolution reconsidered. *Paleobiology*, 3: 115-51.

—— (1986) Punctuated equilibrium at the third stage. *Systematic Zoology*, 35 (1): 143-48.

Gould, S.J., and R. Lewontin (1979) The spandrels of San Marco and the Panglossian paradigm: A critique of the adaptationist programme, *Proceedings of the Royal Society of London*, Series B, 205: 581-98.

Graham, L. (1972) *Science and Philosophy in the Soviet Union* (New York: Knopf).

—— (1981)*Between Science and Values* (New York: Columbia University Press).

—— (1987) *Science, Philosophy, and Human Behavior in the Soviet Union* (New York: Columbia University Press).

Grant, V. (1981) *Plant Speciation*, 2nd ed. (New York: Columbia University Press).

Greene, J. (1971) the Kuhnian paradigm and the Darwinian revolution in natural history. In D. Roller, ed., *Perspectives in the History of Science and Technology* (Norman, OK.: University of Oklahoma Press), 3-25.

Grene, M. (1959) Two evolutionary theories. *British Journal for the Philosophy of Science*, 9: 110-27, 185-93.

—— (1974) *The Understanding of Nature* (Dordrecht, Holland: Reidel).

—— (1978) Sociobiology and the human mind. In M.S. Gregory, A. Silvers, and D. Sutch, eds., *Sociobiology and Human Nature* (San Francisco: Jossey-Bass), 213-24.

—— (1981) Changing concepts of Darwinian evolution. *The Monist*, 64: 195-213.

—— , ed. (1983) *Dimensions of Darwinism* (Cambridge: Cambridge University Press).

—— (1985) Perceptions, interpretation and the sciences: Toward a new philosophy of science. In D.J. Depew and B.H. Weber, eds., *Evolution at a Crossroads* (Cambridge, Mass.: The MIT Press), 1-20.

—— (1986) Philosophy of biology 1983: Problems and prospects. In R.B. Marcus, et al., eds., *Logic, Methodology, and Philosophy of Science* (New York: Elsevier), 7: 433-52.

—— (1987) Hierarchies in biology. *American Scientist*, 75: 504-10.

Griesemer, J.R. (1984) Presentations and the status of theories. In P. Asquith and P. Kitcher, eds., *PSA 1984* (East Lansing, Mich.: Philosophy of Science Association), 1: 102-14.

—— (1988) Causal explanation in laboratory ecology. The case of competitive indeterminancy. In J. Leplin, ed., *PSA 1988* (East Lansing, Mich.: Philosophy of Science Association), 1, forthcoming.

Griesemer, J.R., and M. Wade (1988) Laboratory models, casual explanation and group selection. *Biology and Philosophy*, 3: 67-96.

Griesemer, J.R., and W.C. Wimsatt (1988) Picturing Weismannism: A case study in conceptual evolution. In M. Ruse, ed., *What the Philosophy of Biology Is* (Dordrecht, Holland: Kluwer).

Grobstein, C. *A Double Image of the Double Helix: The Recombinant-DNA Debate* (San Francisco: Freeman).

Hahlweg, K., and C.A. Hooker, eds. (1989) *Issues in Evolutionary Epistemology* (Albany, N.Y.: SUNY Press).

Haila, Y. (1986) On the semiotic dimension of ecological theory: The case of island biogeography. *Biology and Philosophy*, 1: 377-88.

—— (1988) Ecology finding evolution finding ecology. *Biology and Philosophy*, 3, forthcoming.

Haila, Y., and O. Jarrinen (1982) The role of theoretical concepts in understanding the ecological theatre: A case study on island biogeography. In E. Saarinen, ed., *Conceptual Issues in Ecology* (Dordrecht, Holland: Reidel), 261-78.

Hamilton, W.D. (1964a) The genetical evolution of social behaviour. I. *Journal of Theoretical Biology*, 7: 1-16.

—— (1964b) The genetical evolution of social behaviour. II. *Journal of Theoretical Biology,*, 7: 17-32.

—— (1984) Innate social aptitudes of man: An approach from evolutionary genetics. In R.N. Brandon and R.M. Burian, eds., *Genes, Organisms and Populations*, (Cambridge, Mass.: The MIT Press), 193-202.

Haraway, D.J. (1983) The contest for primate nature: Daughters of man-the-hunter in the field, 1960-1980. In M.E. Kann, ed., *The Future of American Democracy: Views from the Left* (Philadelphia: Temple University Press), 175-207.

Harding, S. (1986) *The Science Question in Feminism* (Ithaca, N.Y.: Cornell University Press).

Harding, S., and M.B. Hintikka, eds. (1983) *Discovering Reality: Feminist Perspectives on Epistemology, Metaphysics, Methodology, and Philosophy of Science* (Dordrecht, Holland: Reidel).

Hempel, C.G. (1966) *Philosophy of Natural Science* (Englewood Cliffs, N.J.: Prentice-Hall).

Hennig, W. (1966) *Phylogenetic Systematics* (Urbana: University of Illinois Press).

Hesse, M. (1984) The cognitive claims of metaphor. In J.P. van Noppen, ed., *Metaphor and Religion* (Brussels).

Heyes, C.M., and H.C. Plotkin (1988) Replicators and interactors in cultural evolution. In M. Ruse, ed., *What the Philosophy of Biology Is* (Dordrecht, Holland: Kluwer).

Himmelfarb, G. (1962) *Darwin and the Darwinian Revolution* (New York: Anchor).

Hodge, M.J.S. (1977) The structure and strategy of Darwin's 'long argument'. *British Journal for the History of Science*, 10: 237-46.

—— (1983) The development of Darwin's general biological theorizing. In D.S. Bendall, ed., *Evolution from Molecules to Men* (Cambridge: Cambridge University Press), 43-62.

—— (1988) Darwin's theory and Darwin's argument. In M. Ruse, ed., *What the Philosophy of Biology Is* (Dordrecht, Holland: Kluwer).

Hoffman, A. (1983) Paleobiology at the crossroads: A critique of some modern paleobiological research programs. In M. Grene, ed., *Dimensions of Darwinism* (Cambridge: Cambridge University Press), 241-72.

Holcomb III, H.R. (1987) Criticism, commitment, and the growth of human sociobiology. *Biology and Philosophy*, 2: 43-64.

Holsinger, K.E. (1984) The nature of biological species. *Philosophy of Science*, 51: 293-307.

—— (1987) Discussion: Pluralism and species concepts, or when must we agree with one another? *Philosophy of Science*, 54: 480-85.

Hookway, C. ed.(1984a) *Minds, Machines and Evolution: Philosophical Studies* (Cambridge: Cambridge University Press).

—— (1984b) Naturalism, fallibilism and evolutionary epistemology. In C. Hookway, ed., *Minds, Machines and Evolution:*

Philosophical Studies (Cambridge: Cambridge University Press), 1-16.

Hrdy, S.B. (1981) *The Woman That Never Evolved* (Cambridge: Cambridge University Press).

Hubbard, R. (1983) Have only men evolved? In S. Harding and M.B. Hintikka, eds., *Discovering Reality* (Dordrecht, Holland: Reidel), 45-69.

Hudson, W.D. (1970) *Modern Moral Philosophy* (London: Macmillan). Second edition 1983.

Hughes, W. (1986) Richards' defense of evolutionary ethics. *Biology and Philosophy*, 1: 306-15.

Hull, D.L. (1965) The effect of essentialism on taxonomy: Two thousand years of stasis. *British Journal for the Philosophy of Science*, 15: 314-26, 16: 1-18.

—— (1967) Certainty and circularity in evolutionary taxonomy. *Evolution*, 21: 174-89.

—— (1968) The operational imperative — sense and nonsense in operationism. *Systematic Zoology*, 16: 438-57.

—— (1969) What philosophy of biology is not. *Journal of the History of Biology*, 2: 241-68. Also in *Synthese*, 20: 157-84.

—— (1970a) Contemporary systematic philosophies. *Annual Review of Ecology and Systematics*, 1: 19-54.

—— (1970b) Morphospecies and biospecies: A reply to Ruse. *British Journal for the Philosophy of Science*, 21: 280-82.

—— (1972) Reduction in genetics — biology or philosophy? *Philosophy of Science*, 39: 491-99.

——, ed. (1973a) *Darwin and His Critics* (Cambridge, Mass.: Harvard University Press).

—— (1973b) Reduction in genetics — doing the impossible. In P. Suppes et al., ed., *Logic, Methodology and Philosophy of Science* (The Hague: North-Holland), 4: 619-35.

—— (1974a) *The Philosophy of Biological Science* (Englewood Cliffs, N.J.: Prentice-Hall).

—— (1974b) Are the 'members' of biological species 'similar' to each other? *British Journal for the Philosophy of Science*, 25: 332-34.

—— (1976a) Are species really individuals? *Systematic Zoology*, 25: 174-91.

—— (1976b) Informal aspects of theory reduction. In R.S. Cohen et al., ed., *PSA 1974* (Dordrecht: Reidel), 653-670.

—— (1978a) A matter of individuality. *Philosophy of Science*, 45: 335-60.

—— (1978b) Altruism in science. A sociobiological model of cooperative behaviour among scientists. *Animal Behaviour*, 26: 685-97.

—— (1978c) Sociobiology: Scientific bandwagon or traveling medicine show? *Society*, 15: 50-59. Also in M.S. Gregory, A. Silvers, and D. Sutch, eds., *Sociobiology and Human Nature*, (San Francisco: Jossey-Bass), 136-63.

—— (1979a) Discussion: Reduction in genetics. *Philosophy of Science*, 46: 316-20.

—— (1979b) Philosophy of biology. In P. Asquith and H.E. Kyburg, Jr., eds., *Current Research in Philosophy of Science* (East Lansing, Mich.: Philosophy of Science Association), 421-35.

—— (1979c) The limits of cladism. *Systematic Zoology*, 28: 414-38.

—— (1980) Individuality and selection. *Annual Review of Ecology and Systematics*, 11: 311-32.

—— (1981a) The herd as means. In P. Asquith and R. Giere, eds., *PSA 1980* (East Lansing, Mich.: Philosophy of Science Association), 2: 73-92.

—— (1981b) The principles of biological classification: The use and abuse of philosophy. In P.D. Asquith and I. Hacking, eds., *PSA 1978* (East Lansing, Mich.: Philosophy of Science Association), 2: 130-53.

—— (1982a) The naked meme. In H.C. Plotkin, ed., *Learning, Development and Culture* (Chichester Gt. Britain: J. Wiley and Sons), 273-327.

—— (1982b) Biology and philosophy. In G. Folstad, ed., *Contemporary Philosophy: A New Survey* (The Hague: Nijhoff), 281-316.

—— (1983) Exemplars and scientific change. In P.D. Asquith and T. Nickles, eds., *PSA 1982* (East Lansing, Mich.: Philosophy of Science Association), 2: 479-503.

—— (1984a) Units of evolution: A metaphysical essay. In R.N. Brandon and R.M. Burian, eds., *Genes, Organisms and Populations* (Cambridge, Mass.: The MIT Press), 142-60.

—— (1984b) Historical entities and historial narratives. In C. Hookway, ed., *Minds, Machines and Evolution* (Cambridge: Cambridge University Press), 17-42.

—— (1984c) Cladistic theory: Hypotheses that blur and grow. In T. Duncan and T. Stuessy, eds., *Cladistic Perspectives on the Reconstruction of Evolutionary History* (New York: Columbia University Press), 5-23.

—— (1985a) Bias and commitment in science. Phenetics and cladistics. *Annals of Science*, 42: 319-38.

—— (1985b) Darwinism as a historical entity: A historiographic proposal. In D. Kohn, ed., *The Darwinian Heritage* (Princeton, N.J.: Princeton University Press), 773-812.

—— (1986) Conceptual evolution and the eye of the octopus. In R.B. Marcus, G.J.W. Dorn, and P. Weingartner, ed., *Logic, Methodology and Philosophy of Science* (Amsterdam: North-Holland), 7: 643-65.

—— (1987) Genealogical actors in ecological roles. *Biology and Philosophy*, 2: 168-83.

—— (1988a) *Science as a Process: An Evolutionary Account of the Social and Conceptual Development of Science* (Chicago: University of Chicago Press).

—— (1988b) A mechanism and its metaphysics: An evolutionary account of the social and conceptual development of science. *Biology and Philosophy*, 3: 123-55.

Hume, D. (1978) *A Treatise of Human Nature* (Oxford: Oxford University Press). First published 1739-40.

Hunkapiller, T., et al. (1982) The impact of modern genetics on evolutionary theory. In R. Milkman, ed., *Perspectives on Evolution* (Sunderland, Mass.: Sinauer), 164-89.

Huxley, J. ed. (1947) *Evolution and Ethics* (London: Pilot).

Isaac, G. (1983) Aspects of human evolution. In D.S. Bendall, ed., *Evolution from Molecules to Men* (Cambridge: Cambridge University Press), 509-43.

Jacobs, J. (1986) Teleology and reduction in biology. *Biology and Philosophy*, 1: 389-400.

Janvier, P. (1984) Cladistics: Theory, purpose and evolutionary implications. In J. Pollard, ed., *Evolutionary Theory: Paths into the Future* (Chichester, Gt. Britain: J. Wiley and Sons), 39-76.

Johanson, D., and M. Edey (1981) *Lucy: The Beginnings of Humankind* (New York: Simon and Schuster).

Jones, G. (1980) *Social Darwinism and English Thought* (Brighton, Gt. Britain: Harvester).

Jongeling, T.B. (1985) On an axiomatization of evolutionary theory. *Journal of Theoretical Biology*, 117: 529-43.

Jungck, J.R. (1983) Is the neodarwinian synthesis robust enough to withstand the challenge of recent discoveries in molecular biology and molecular evolution? In P. Asquith and T. Nickles, eds., *PSA 1982* (East Lansing, Mich.: Philosophy of Science Association), 2: 322-30.

Kant, I. (1929) *Critique of Pure Reason*, trans. N. Kemp-Smith (London: Macmillan). First published 1781.

Kary, C. (1983) Can Darwinian inheritance be extended from biology to epistemology? In P. Asquith and T. Nickles, eds., *PSA 1982* (East Lansing, Mich.: Philosophy of Science Association), 2: 356-69.

Kauffman, S.A. (1977) Constraints on the sociobiologists' program. In F. Suppe and P. Asquith, eds., *PSA 1976* (East Lansing, Mich.: Philosophy of Science Association), 2: 32-47.

—— (1983) Filling some epistemological gaps: New patterns of inference in evolutionary theory. In P. Asquith and T. Nickles, eds., *PSA 1982* (East Lansing, Mich.: Philosophy of Science Association), 2: 292-313.

—— (1985) Self-organization, selective adaptation, and its limits: A new pattern of inference in evolution and development. In D.J. Depew and B.H. Weber, eds., *Evolution at a Crossroads* (Cambridge, Mass.: The MIT Press), 169-208.

—— (1986) A framework to think about evolving genetic regulatory systems. In W. Bechtel, ed., *Integrating Scientific Disciplines* (Dordrecht, Holland: Nijhoff), 149-64.

Kawata, M. (1987) Units and passages: A view for evolutionary biology and ecology. *Biology and Philosophy*, 2: 415-34.

Keller, E.F. (1983) *A Feeling For the Organism* (San Francisco: Freeman).

—— (1984) *Reflections on Gender and Science* (New Haven, Conn.: Yale University Press).

—— (1987) Reproduction and the central project of evolutionary theory. *Biology and Philosophy*, 2: 383-96.

Kellogg, D. (1988) "And then a miracle occurs" — weak links in the chain of argument from punctuation to hierarchy. *Biology and Philosophy*, 3: 3-28.

Kimbrough, S.O. (1979) On the reduction of genetics to molecular biology. *Philosophy of Science*, 46: 389-406.

Kimura, M. (1983) *The Neutral Theory of Molecular Evolution* (Cambridge: Cambridge University Press).

Kincaid, H. (1987) Supervenience doesn't entail reducibility. *The Southern Journal of Philosophy*, 25: 343-56.

Kitcher, P. (1982) Genes. *British Journal for the Philosophy of Science*, 33: 337-59.

—— (1983) *Abusing Science* (Cambridge, Mass.: The MIT Press).

—— (1984a) 1953 and all that: A tale of two sciences. *Philosophical Review*, 93: 335-73.

—— (1984b) Species. *Philosophy of Science*, 51: 308-35.

—— (1984c) Against the monism of the moment: A reply to Elliott Sober. *Philosophy of Science*, 51: 616-30.

—— (1985a) *Vaulting Ambition* (Cambridge, Mass.: The MIT Press).

—— (1985b) Darwin's achievement. In N. Rescher, ed., *Reason and Rationality in Science*, (Washington, D.C.: University Press of America), 127-89.

—— (1987a) Ghostly whispers: Mayr, Ghiselin, and the "philosophers" on the ontological status of species. *Biology and Philosophy*, 2: 184-91.

—— (1987b) Why not the best? In J. Dupré, ed., *The Latest on the Best: Essays on Evolution and Optimality*, (Cambridge, Mass.: The MIT Press), 77-102.

—— (1987c) Précis of Vaulting Ambition: Sociobiology and the quest for human nature. *Behavorial and Brain Sciences,* 10: 61-99. [This includes commentary by 22 respondents and author's response.]

—— (1988) Some puzzles about species. In M. Ruse, ed., *What the Philosophy of Biology Is* (Dordrecht, Holland: Kluwer).

Kitts, D.B. (1987) Plato on kinds of animals. *Biology and Philosophy,* 2: 315-28.

Kitts, D.B., and D.J. Kitts (1979) Biological species as natural kinds. *Philosophy of Science,* 46: 613-22.

Kleiner, S.A. (1985) Darwin's and Wallace's revolutionary research programme. *British Journal for the Philosophy of Science,* 36: 367-92.

—— (1988) Darwin's logic of discovery: The search for a mechanism before Malthus. *Biology and Philosophy,* 3: forthcoming.

Kohn, D., ed. (1985) *The Darwinian Heritage* (Princeton, N.J.: Princeton University Press).

Kuhn, T.S. (1962) *The Structure of Scientific Revolutions* (Chicago: University of Chicago Press).

—— (1979) Metaphor in science. In A. Ortony, ed., *Metaphor and Thought* (Cambridge, Mass.: Cambridge University Press), 409- 19.

Kyburg, H.E. (1968) *Philosophy of Science: A Formal Approach* (New York: Macmillan).

Lakoff, G., and M. Johnson (1980) *Metaphors We Live By* (Chicago: University of Chicago Press).

Landau, M. (1984) Human evolution as narrative. *American Scientist,* 72 (3): 262-68.

Laudan, L. (1982) Science at the bar: Causes for concern. In J. Murphy, *Evolution, Morality and the Meaning of Life* (Totowa, N.J.: Rowman and Littlefield), 149-54. (Also in *Science, Technology, and Human Values,* (1982) 7: 16-19.)

Leeds, A., and V. Dusek, eds. (1981) *Sociobiology: The Debate Evolves.* In *Philosophical Forum,* 2 (3): 1-323.

Levins, R., and R. Lewontin (1985) *The Dialectical Biologist* (Cambridge, Mass.: Harvard University Press).

Lewontin, R.C. (1969) The bases of conflict in biological explanation. *Journal of the History of Biology*, 2: 35-46.

—— (1974) *The Genetic Basis of Evolutionary Change* (New York: Columbia University Press).

—— (1977) Sociobiology: A caricature of Darwinism. In P. Asquith and F. Suppe, eds., *PSA 1976* (East Lansing, Mich.: Philosophy of Science Association), 2: 22-31.

—— (1978) Adaptation. *Scientific American*, 239 (3): 212-30.

—— (1982) Organism and environment. In H.C. Plotkin, ed., *Learning, Development, and Culture: Essays in Evolutionary Epistemology* (Chichester, Gt. Britain: J. Wiley and Sons), 151-70.

Lewontin, R., and R. Levins (1976) The problem of Lysenkoism. In H. and S. Rose, eds., *The Radicalisation of Science* (London: Macmillan), 32-64.

Lewontin, R.C., S. Rose, and L. Kamin (1984) *Not in our Genes: Biology, Ideology, and Human Nature* (New York: Pantheon).

Lieberman, P. (1984) *The Biology and Evolution of Language* (Cambridge, Mass.: Harvard University Press).

Lindenmayer, A. (1973) Cellular automata, formal languages and developmental systems. In P. Suppes et al., eds., *Logic, Methodology and Philosophy of Science*, (The Hague: North Holland).

—— (1975) Developmental algorithms for multicellular organisms. *Journal of Theoretical Biology*, 54: 3-22.

Lloyd, E.A. (1983) The nature of Darwin's support for the theory of natural selection. *Philosophy of Science*, 50: 112-29.

—— (1986a) Thinking about models in evolutionary theory. *Philosophica*, 37: 87-100.

—— (1986b) Evaluation of evidence in group selection debates. In A. Fine and P. Kitcher, eds., *PSA 1986* (East Lansing, Mich.: Philosophy of Science Association), 1: 483-93.

—— (1986c) Mathematical models and the structure of evolutionary theory. In P. Weingartner and G. Dorn, eds., *Foundations of Biology* (Vienna: Holder-Pichler-Tempsky).

—— (1987a) Confirmation of ecological and evolutionary models. *Biology and Philosophy*, 2: 277-94.

—— (1987b) Response to Sloep and van der Steen. *Biology and Philosophy*, 2: 23-25.

—— (1988a) A structural approach to defining units of selection. *Philosophy of Science*, forthcoming.

—— (1988b) *The Structure and Confirmation of Evolutionary Theory.* (Westport, Conn.: Greenwood Press).

Locke, J. (1959) *An Essay Concerning Human Understanding*, ed. A.C. Fraser (New York: Dover). First published 1690.

Lorenz, K. (1941) Kant's Lehre von a priorischen im Lichte geganwärtiger Biologie. *Blätter für Deutsche Philosophie*, 15: 94-125. Translated and reprinted as: Kant's doctrine of the a priori in the light of contemporary biology, in H.C. Plotkin, ed., *Learning, Development, and Culture: Essays in Evolutionary Epistemology* (Chichester, Gt. Britain: J. Wiley and Sons, 1982), 121-43.

Losee, J. (1972) *A Historical Introduction to the Philosophy of Science* (Oxford: Oxford University Press).

Lovtrup, S. (1987) *Darwinism: The Refutation of a Myth* (Beckenham, Gt. Britain: Croom Helm).

Lucas, J.R. (1979) Wilberforce and Huxley: A legendary encounter *Historical Journal*, 22: 313-30.

Lumsden, C., and E.O. Wilson (1981) *Genes, Mind, and Culture* (Cambridge, Mass.: Harvard University Press).

—— (1983) *Promethean Fire* (Cambridge, Mass.: Harvard University Press).

Mackie, J.L. (1966) The direction of causation. *Philosophical Review*, 75: 441-66.

—— (1978) The law of the jungle. *Philosophy*, 53: 553-73.

—— (1980) *Hume's Moral Theory* (London: Routledge and Kegan Paul).

Manser, A.R. (1965) The concept of evolution. *Philosophy*, 40: 18-34.

Matthen, M. (1988) Biological functions and perceptual content. *Journal of Philosophy*, 85: 5-27.

Matthen, M., and E. Levy (1984) Teleology, error, and the human immune system. *Journal of Philosophy*, 81: 351-72.

—— (1986) Organic teleology. In N. Rescher, ed., *Current Issues in Teleology* (Landham, Md.: University Press of America), 93-101.

Maull, N. (1977) Unifying science without reduction. *Studies in History and Philosophy of Science*, 8: 143-62.

Maynard Smith, J. (1974) The theory of games and the evolution of animal conflict. *Journal of Theoretical Biology*, 47: 209-21.

—— (1981) Did Darwin get it right? *London Review of Books*, 3 (11): 10-11.

—— (1983a) *Evolution and the Theory of Games* (Cambridge: Cambridge University Press).

—— (1983b) Current controversies in evolutionary biology. In M. Grene, ed., *Dimensions of Darwinism* (Cambridge: Cambridge University Press), 273-88.

—— (1984a) Group selection. In R.N. Brandon and R.M. Burian, eds., *Genes, Organisms and Populations* (Cambridge, Mass.: The MIT Press), 238-49.

—— (1984b) The evolution of animal intelligence. In C. Hookway, ed., *Minds, Machines and Evolution* (Cambridge: Cambridge University Press), 63-72.

—— (1987) Darwinism stays unpunctured. *Nature*, 330: 516.

Maynard Smith, J., R. Burian, S. Kauffman, P. Alberch, J. Campbell, B. Goodwin, R. Lande, D. Raup, and L. Wolpert (1985) Developmental constraints and evolution. *Quarterly Review of Biology*, 60: 265-87.

Mayr, E. (1942) *Systematics and the Origin of Species* (New York: Columbia University Press).

—— , ed. (1957) *The Species Problem* (Washington, D.C.: A.A.A.S.), Publication 50.

—— (1963) *Animal Species and Evolution* (Cambridge, Mass.: Belknap).

—— (1969a) *Principles of Systematic Zoology* (New York: McGraw-Hill).

—— (1969b) Scientific explanation and conceptual framework. *Journal of the History of Biology*, 2: 123-28.

—— (1971) The nature of the Darwinian Revolution. *Science*, 176: 981-89.

—— (1974) Teleological and teleonomic, a new analysis. In R.S. Cohen and M.W. Wartofsky, eds., *Boston Studies in the Philosophy of Science* (Dordrecht, Holland: Reidel), 14: 91-117.

—— (1975) The unity of the genotype. *Biologisches Zentralblatt*, 94: 377-88.

—— (1976) Is the species a class or an individual? *Systematic Zoology*, 25: 192.

—— (1981) Biological classification: Toward a synthesis of opposing methodologies. *Science*, 214: 510-16.

—— (1982) *The Growth of Biological Thought: Diversity, Evolution, and Inheritance* (Cambridge, Mass.: Harvard University Press).

—— (1983) Comments on David Hull's paper on exemplars and type specimens. In P. Asquith and T. Nickles, eds., *PSA 1982* (East Lansing, Mich.: Philosophy of Science Association), 2: 504-11.

—— (1984) The unity of genotype. In R.N. Brandon and R.M. Burian, eds., *Genes, Organisms and Populations* (Cambridge, Mass.: The MIT Press), 69-85.

—— (1985a) What is Darwinism today? In P. Asquith and P. Kitcher, eds., *PSA 1984* (East Lansing, Mich.: Philosophy of Science Association), 2: 145-56.

—— (1985b) How biology differs from the physical sciences. In D.J. Depew and B.H. Weber, eds., *Evolution at a Crossroads* (Cambridge, Mass.: The MIT Press), 43-64.

—— (1987) The ontological status of species: Scientific progress and philosophical terminology. *Biology and Philosophy*, 2: 145-66.

—— (1988) A response to David Kitts, *Biology and Philosophy*, 3: 97-8.

McMullin, E., ed. (1986) *The Creation-Evolution Controversy* (Notre Dame, Ind.: University of Notre Dame Press).

Michod, R.E. (1981) Positive heuristics in evolutionary biology. *British Journal for the Philosophy of Science*, 32: 1-36.

—— (1984) The theory of kin selection. In R.N. Brandon and R.M. Burian, eds., *Genes, Organisms and Populations* (Cambridge, Mass.: The MIT Press), 203-37.

Midgley, M. (1978) *Beast and Man: The Roots of Human Nature* (Ithaca, N.Y.: Cornell University Press).

—— (1979) Gene-juggling. *Philosophy*, 54: 439-58.

—— (1985) *Evolution as a Religion: Strange Hopes and Stranger Fears* (London: Methuen).

Milkman, R. (1982) Toward a unified selection theory. In R. Milkman, ed., *Perspectives on Evolution* (Sunderland, Mass.: Sinauer), 105-18.

Mills, S.K., and J.H. Beatty (1979) The propensity interpretation of fitness. *Philosophy of Science*, 46: 263-86.

Mishler, B.D., and R.N. Brandon (1987) Individuality, pluralism, and the phylogenetic species concept. *Biology and Philosophy*, 2: 397-414.

Mishler, B.D., and M.J. Donoghue (1982) Species concepts: A case for pluralism. *Systematic Zoology*, 31: 503-11.

Mitchell, S.D. (1987) Competing units of selection? A case of symbiosis. *Philosophy of Science*, 54: 351-67.

Mlikovsky, J., and V. Novak, eds. (1985) *Evolution and Morphogensis* (Prague: Czechoslovak Academy of Sciences).

Montagu, A., ed. (1980) *Sociobiology Examined* (Oxford: Oxford University Press).

—— , ed. (1984) *Science and Creationism* (Oxford: Oxford University Press).

Moore, J.R. (1987) Born again Social Darwinism. *Annals of Science*, 44: 409-17.

Morowitz, H. (1986) Entropy and nonsense: A review of Daniel R. Brooks and E.O. Wiley, *Evolution as Entropy*. *Biology and Philosophy*, 1: 473-76.

—— (1987) The mind body problem and the second law of thermodynamics. *Biology and Philosophy*, 2: 271-76.

Morris, H.M., ed. (1974) *Scientific Creationism* (San Diego, Calif.: Creation-Life Publishers).

Mortensen, V. and R.C. Sorensen, eds. (1987) *Free Will and Determinism* (Aarhus, Denmark: Aarhus University Press).

Muller, H.J. (1949) The Darwinian and modern conceptions of natural selection. *Proceedings of the American Philosophical Society*, 93: 459-70.

Munson, R., ed. (1971) *Man and Nature: Philosophical Issues in Biology* (New York: Delta Books).

—— (1975) Is biology a provincial science? *Philosophy of Science*, 42: 428-47.

Murphy, J.G. (1982) *Evolution, Morality, and the Meaning of Life* (Totowa, N.J.: Rowman and Littlefield).

Nagel, E. (1961) *The Structure of Science* (New York: Harcourt, Brace, and World).

—— (1977) Teleology revisited. *Journal of Philosophy*, 74: 261-301.

Nagel, T. (1986) *The View from Nowhere* (Oxford: Oxford University Press).

Nelson, G. (1973) Classification as an expression of phylogenetic relationship. *Systematic Zoology*, 22: 344-59.

Nelson, G., and N. Platnick (1981) *Systematics and Biogeography: Cladistics and Vicariance* (New York: Columbia University Press).

Nissen, L. (1983) Discussion: Wright on teleological descriptions of goal-directed behaviour. *Philosophy of Science*, 50: 151-57.

Nitecki, M.H., ed. (1988) *The Idea of Progress in Evolution* (Chicago: University of Chicago Press).

Nozick, R. (1981) *Philosophical Explanations* (Cambridge, Mass.: Harvard University Press).

O'Grady, R.T. (1986) Historical processes, evolutionary explanations, and problems with teleology. *Canadian Journal of Zoology*, 64: 1010-20.

Olding, A. (1978) A defence of evolutionary laws. *British Journal for the Philosophy of Science*, 29: 131-43.

Oldroyd, D.R. (1986) Charles Darwin's theory of evolution: A review of our present understanding. *Biology and Philosophy*, 1: 133-68.

Ospovat, D. (1981) *The Development of Darwin's Theory* (Cambridge: Cambridge University Press).

Oster, G.F., and E.O. Wilson (1978). *Caste and Ecology in the Social Insects* (Princeton, N.J.: Princeton University Press).

Overton, W.R. (1982) Creationism in schools: The decision in McLean versus the Arkansas Board of Education. *Science*, 215: 934-43.

Paley, W. (1802) *Natural Theology*. In *Collected Works* (London: Rivington, 1819), vol. 4.

Paterson, H.E.H. (1978) More evidence against speciation by reinforcement. *South African Journal of Science*, 74: 369-71.

—— (1980) A consensus on "mate recognition systems." *Evolution*, 34: 330-31.

Patterson, C. (1978a) Verifiability in systematics. *Systematic Zoology*, 27: 218-21.

—— (1978b) *Evolution* (London: British Museum, Natural History).

—— (1981) The significance of fossils in determining evolutionary relationships. *Annual Review of Ecology and Systematics*, 12: 195-223.

—— (1982) Classes and cladists or individuals and evolution. *Systematic Zoology*, 31: 284-86.

—— (1983) How does phylogeny differ from ontogeny? In B.C. Goodwin, N. Holder, and C.C. Wylie, eds., *Development and Evolution* (Cambridge: Cambridge University Press), 1-31.

Peacocke, A. (1986) *God and the New Biology* (London: Dent).

Peters, D.S. (1983) Evolutionary theory and its consequences for the concept of adaptation. In M. Grene, ed., *Dimensions of Darwinism* (Cambridge: Cambridge University Press), 315-27.

Peters, R.H. (1976) Tautology in evolution and ecology. *American Naturalist*, 110: 1-12.

Pilbeam, D. (1984) The descent of hominoids and hominids. *Scientific American*, 250 (3), March: 87-97.

Platnick, N. (1982) Defining characters and evolutionary groups. *Systematic Zoology*, 31: 282-84.

Platnick, N., and E. Gaffney (1978) Evolutionary biology: A Popperian perspective. *Systematic Zoology*, 27: 137-41.

Platnick, N.I., and G. Nelson (1981) The purpose of biological classification. In P. Asquith and I. Hacking, eds., *PSA 1978* (East Lansing, Mich.: Philosophy of Science Association), 2: 117-29.

Plotkin, H.C., ed., (1982) *Learning, Development, and Culture: Essays in Evolutionary Epistemology* (Chichester, Gt. Britain: J. Wiley and Sons).

—— (1987) Evolutionary epistemology as science. *Biology and Philosophy*, 2: 295-314.

Pollard, J., ed. (1984) *Evolutionary Theory: Paths into the Future* (Chichester, Gt. Britain: J. Wiley and Sons).

Popper, K.R. (1959) *The Logic of Scientific Discovery* (London: Hutchinson).

—— (1963) *Conjectures and Refutations* (London: Routledge and Kegan Paul).

—— (1972) *Objective Knowledge* (Oxford: Oxford University Press).

—— (1974) Intellectual autobiography. In P.A. Schilpp, ed., *The Philosophy of Karl Popper* (LaSalle, Ill.: Open Court) 1: 3-181.

—— (1978) Natural selection and the emergence of mind. *Dialectica*, 32: 339-55.

—— (1980) Letter to the editor. *New Scientist*, 87: 611.

—— (1984) Evolutionary epistomology. In J. Pollard, ed., *Evolutionary Theory: Paths into the Future* (Chichester, Gt. Britain: J. Wiley and Sons), 239-56.

Pratt, V.J.F. (1972) Biological classification, *British Journal for the Philosophy of Science*, 23: 305-27.

Provine, W.B. (1983) The development of Wright's theory of evolution: Systematics, adaptation, and drift. In M. Grene, ed., *Dimensions of Darwinism* (Cambridge: Cambridge University Press), 43-70.

Putnam, H. (1982) Why reason can't be naturalized. *Synthese*, 52: 3-23.

Quine, W.V.O. (1953) Two dogmas of empiricism. In *From a Logical Point of View* (Cambridge, Mass.: Harvard University Press), 20-46.

—— (1969a) Epistemology naturalized. In *Ontological Relativity and Other Essays* (New York: Columbia University Press), 69-90.

—— (1969b) Natural kinds. In *Ontological Relativity and Other Essays*. (New York: Columbia University Press), 114-138.

Quinn, P. (1984) The philosopher of science as expert witness. In J.T. Cushing, et al., eds., *Science and Reality* (Notre Dame, Ind.: University of Notre Dame Press), 32-53.

Radnitzky, G., and W. Bartley, eds. (1987) *Evolutionary Epistemology, Theory of Rationality, and the Sociology of Knowledge* (La Salle, Ill.: Open Court).

Recker, D.A. (1987) Causal efficacy: The structure of Darwin's argument strategy in the 'Origin of Species'. *Philosophy of Science*, 54: 147-75.

Reed, Edward (1978) Group selection and methodological individualism: A criticism of Watkins. *British Journal for the Philosophy of Science*, 29: 256-62.

Reed, Evelyn (1975) *Woman's Evolution* (New York: Pathfinder Press).

Regan, T. (1983) *The Case for Animal Rights* (Berkeley, Calif.: University of California Press).

Regis, E., ed. (1985) *Extraterrestrials: Science and Alien Intelligence* (Cambridge: Cambridge University Press).

Reif, W.E. (1983) Evolutionary theory in German paleontology. In M. Grene, ed., *Dimensions of Darwinism* (Cambridge: Cambridge University Press), 173-204.

Rescher, N., ed. (1986) *Current Issues in Teleology* (Lanham, Md.: University Press of America).

Reynolds, V., and R. Tanner (1983) *The Biology of Religion* (London: Longman).

Rhodes, F.H.T. (1986) Darwinian gradualism and its limits. The development of Darwin's views on the rate and pattern of evolutionary change. *Journal of the History of Biology*, 20: 139-58.

Richards, J., ed. (1978) *Recombinant DNA: Science, Ethics, and Politics* (New York: Academic Press).

Richards, R.J. (1977) Discussion: The natural selection model of conceptual evolution. *Philosophy of Science*, 44, 494-501.

—— (1986a) A defense of evolutionary ethics. *Biology and Philosophy*, 1: 265-92.

—— (1986b) Justification through biological faith: A rejoinder. *Biology and Philosophy*, 1: 337-54.

—— (1987) *Darwin and the Emergence of Evolutionary Theories of Mind and Behavior* (Chicago: University of Chicago Press).

Richardson, R.C. (1983) Grades of organization and the units of selection controversy. In P. Asquith and T. Nickles, eds., *PSA 1982* (East Lansing, Mich.: Philosophy of Science Association), 2: 324-40.

—— (1986a) Models and scientific explanations. *Philosophica*, 37: 59-72.

—— (1986b) Language, thought, and communication. In W. Bechtel, ed., *Integrating Scientific Disciplines*. (Dordrecht, Holland: Nijhoff), 263-84.

—— (1986c) Models and scientific idealizations. In P. Weingartner and G. Dorn, eds., *Foundations of Biology* (Vienna: Holder-Pichler-Tempsky), 109-44.

Riddiford, A., and D. Penny (1984) The scientific status of modern evolutionary theory. In J. Pollard, ed., *Evolutionary Theory: Paths into the Future* (Chichester, Gt. Britain: J. Wiley and Sons), 1-38.

Ridley, M. (1985) *The Problems of Evolution* (Oxford: Oxford University Press).

—— (1986) *Evolution and Classification: The Reformation of Cladism* (New York: Longman).

Riedl, R. (1980) *Biologie der Erkenntnis.* Written in collaboration with R. Kaspar (Berlin-Hamburg: P. Parey). Translated as *Biology of Knowledge* (Chichester, Gt. Britain: J. Wiley and Sons, 1984).

—— (1983) The role of morphology in the theory of evolution. In M. Grene, ed., *Dimensions of Darwinism* (Cambridge: Cambridge University Press), 205-38.

Rieppel, O. (1986) Species are individuals: A review and critique of the argument. In M. Hecht, B. Wallace, and G. Prance, eds., *Evolutionary Biology* (New York: Plenum), 20: 283-317.

—— (1988) Louis Agassiz (1807-1873) and the reality of natural groups. *Biology and Philosophy,* 3: 29-48.

Rizzotti, M. and Zanardo, A. (1986) Axiomatization of genetics 1. Biological meaning. *Journal of Theoretical Biology,* 118: 61-71.

Robinson, J.D. (1986) Reduction, explanation and the quest of biological reasearch. *Philosophy of Science,* 53: 333-53.

Rolston, H. (1986) *Philosophy Gone Wild* (Buffalo, N.Y.: Prometheus).

Rose H., and S., eds. (1976) *The Radicalisation of Science* (London, Macmillian).

——, eds. (1982) *Towards a Liberatory Biology* (London: Allison and Busby).

Rosen, D.E. (1984) Hierarchies and history. In J. Pollard, ed., *Evolutionary Theory: Paths into the Future* (Chichester, Gt. Britain: J. Wiley and Sons), 77-98.

Rosenberg, A. (1978) The supervenience of biological concepts. *Philosophy of Science,* 46: 263-86.

—— (1980a) Ruse's treatment of the evidence for evolution: A reconsideration. In P. Asquith and R. Giere, eds., *PSA 1980* (East Lansing, Mich.: Philosophy of Science Association), 1: 83-93.

—— (1980b) Species notions and the theoretical hierarchy of biology. *Nature and System,* 2: 163-72.

—— (1981) *Sociobiology and the Preemption of Social Science* (Baltimore, Md.: Johns Hopkins University Press).

—— (1982a) Discussion: On the propensity definition of fitness. *Philosophy of Science,* 49: 268-73.

—— (1982b) Causation and teleology. In G. Folstad, ed., *Contemporary Philosophy: A New Survey* (The Hague: Nijhoff), 51-86.

—— (1983a) Discussion: Coefficients, effects and genetic selection. *Philosophy of Science,* 50: 332-38.

—— (1983b) Fitness. *Journal of Philosophy,* 80: 457-73.

—— (1983c) Critical notice of *Genes, Mind, and Culture. Journal of Philosophy,* 80: 304-11.

—— (1985a) *The Structure of Biological Science* (Cambridge: Cambridge University Press).

—— (1985b) Adaptationist imperatives and panglossian paradigms. In J. Fetzer, ed., *Sociobiology and Epistemology* (Dordrecht, Holland: Reidel), 161-80.

—— (1985c) Darwinism today — and tomorrow, but not yesterday. In P. Asquith and P. Kitcher, eds., *PSA 1984* (East Lansing, Mich.: Philosophy of Science Association), 2: 157-78.

—— (1986a) Causation and explanation in evolutionary theory. *Behaviourism,* 14: 77-88.

—— (1986b) Ignorance and disinformation in the philosophy of biology. A reply to Stent. *Biology and Philosophy,* 1: 461-72.

—— (1986c) Intention and action amongst the macromolecules. In N. Rescher, ed., *Current Issues in Teleology* (Landham, Md.: University Press of America), 65-76.

—— (1987) Why do the nature of species matter? Comments on Ghiselin and Mayr. *Biology and Philosophy,* 2: 192-97.

—— (1988a) From reductionsim to instrumentalism, or qualms of a former reductionist. In M. Ruse, ed., *What the Philosophy of Biology Is* (Dordrecht, Holland: Kluwer).

—— (1988b) Is the theory of natural selection really a statistical theory? *Canadian Journal of Philosophy,* forthcoming.

Rosenberg, A., and M.B. Williams (1985) Fitness in fact and fiction. *Journal of Philosophy,* 82: 738-49.

—— (1986) Discussion: Fitness as primitive and propensity. *Philosophy of Science,* 53: 416-18.

Rosenblueth, A., N. Wiener, and J. Bigelow (1943) Behavior, purpose and teleology. *Philosophy of Science,* 10: 18-25.

Rudwick, M. (1986) *The Great Devonian Controversy* (Chicago: University of Chicago Press).

Rumbaugh, D.M. (1986) From genes to genius in the quest for control. In W. Bechtel, ed., *Integrating Scientific Disciplines* (Dordrecht, Holland: Nijhoff), 297-308.

Ruse, M. (1969a) Definitions of species in biology. *The British Journal for the Philosophy of Science*, 20: 97-119.

—— (1969b) Confirmation and falsification of theories of evolution. *Scientia*, 104: 329-57.

—— (1970) The revolution in biology. *Theoria*, 36: 1-22.

—— (1971a) Functional statements in biology. *Philosophy of Science*, 38: 87-95.

—— (1971b) Natural selection in *The Origin of Species*. *Studies in History and Philosophy of Science*, 1: 311-510.

—— (1971c) The species problem: A reply to Hull. *British Journal for the Philosophy of Science*, 22: 369-71.

—— (1972a) Is the theory of evolution different? I. The central core of the theory. *Scientia*, 106: 765-83.

—— (1972b) Is the theory of evolution different? II. The structure of the entire theory. *Scientia*, 106: 1069-93.

—— (1972c) Biological adaptation. *Philosophy of Science*, 39: 525-28.

—— (1973a) *The Philosophy of Biology* (London: Hutchinson).

—— (1973b) Teleological explanation and the animal world. *Mind*, 82: 433-36.

—— (1973c) A reply to Wright's analysis of functional statements. *Philosophy of Science*, 40: 277-80.

—— (1975a) Woodger on genetics: A critical evaluation. *Acta Biotheoretica*, 24 (1-2): 1-13.

—— (1975b) Darwin's debt to philosophy: An examination of the influence of the philosophical ideas of John F.W. Hershel and William Whewell on the development of Charles Darwin's theory of evolution. *Studies in History and Philosophy of Science*, 6: 159-81.

—— (1976a) Reduction in genetics. In R.S. Cohen et al., eds., *PSA 1974* (Dordrecht, Holland: Reidel), 633-52.

—— (1976b) The Scientific methodology of William Whewell. *Centaurus*, 20: 227-57.

—— (1977a) Karl Popper's philosophy of biology. *Philosophy of Science*, 44: 638-61.

—— (1977b) Is biology different from physics? In R. Colodny, ed., *Logic, Laws, and Life* (Pittsburgh, Pa.: University of Pittsburgh Press), 89-127.

—— (1977c) Sociobiology: Sound science or muddled metaphysics? In F. Suppe and P. Asquith, eds., *PSA 1976* (East Lansing, Mich.: Philosophy of Science Association), 2: 48-73.

—— (1979a) *The Darwinian Revolution: Science Red in Tooth and Claw* (Chicago: University of Chicago Press).

—— (1979b) *Sociobiology: Sense or Nonsense?* (Dordrecht, Holland; Reidel). Second edition, 1985.

—— (1980) Charles Darwin and group selection. *Annals of Science*, 37: 615-30. Reprinted in R.N. Brandon and R. M. Burian, eds., *Genes, Organisms and Populations* (Cambridge, Mass.: The MIT Press, 1984), 9-28

—— (1981a) Are there gay genes? Sociobiology looks at homosexuality. *Journal of Homosexuality*, 6 (4): 5-34.

—— (1981b) Darwin's theory: an exercise in science. *New Scientist*, 90 (25th June): 828-30.

—— (1981c) The last word on teleology, or optimality models vindicated. In *Is Science Sexist? And Other Essays on the Biomedical Sciences* (Dordrecht, Holland: Reidel), 85-101.

—— (1981d) Is human sociobiology a new paradigm? *Philosophical Forum*, 13 (2/3): 119-43.

—— (1982a) *Darwinism Defended: A Guide to the Evolution Controversies* (Reading, Mass.: Addison-Wesley).

—— (1982b) The Arkansas creation trial 1981: Is there a message for us all? *The History and Social Science Teacher*, 18: 23-28.

—— (1982c) A philosopher at the monkey trial. *New Scientist*, 93: 317-19.

—— (1984a) A philosopher's day in court. In A. Montagu, ed., *Science and Creationism* (New York: Oxford University Press), 311-42.

—— (1984b) Critical Notice: Philip Kitcher's "Abusing science: The case against creationism" *Philosophy of Science*, 51: 348-54.

—— (1984c) Is there a limit to our knowledge of evolution? *BioScience*, 34 (2): 100-104.

—— (1985a) Evolutionary epistemology: Does sociobiology help? In J. Fetzer, ed., *Sociobiology and Epistemology* (Dordrecht, Holland: Reidel), 249-265.

—— (1985b) Biological science and feminist values. In P. Asquith and P. Kitcher, eds., *PSA 1984* (East Lansing, Mich.: Philosophy of Science Association), 2: 525-42.

—— (1986a) Teleology and the biological sciences. In N. Rescher, ed., *Current Issues in Teleology* (Lanham, Md.: University Press of America), 56-64.

—— (1986b) Evolutionary ethics: A phoenix arisen. *Zygon,* 21: 95-112.

—— (1986c) *Taking Darwin Seriously: A Naturalistic Approach to Philosophy* (Oxford: Blackwell).

—— (1987a) Darwinism and determinism. *Zygon,* 22: 419-42.

—— (1987b) Species: Natural kinds, individuals, or what? *British Journal for the Philosophy of Science,* 38: 225-42.

—— (1987c) Is sociobiology a new paradigm? *Philosophy of Science,* 54: 98-104.

—— , ed. (1988a) *But is it Science? The Philosophical Question in the Creation/Evolution Controversy* (Buffalo, N.Y.: Prometheus).

—— , ed. (1988b) *What the Philosophy of Biology Is* (Dordrecht, Holland: Kluwer).

—— , ed. (1988c) *Readings in the Philosophy of Biology* (New York: Macmillan).

—— (1988d) *Homosexuality: A Philosophical Inquiry* (Oxford: Blackwell).

—— (1988e) *The Darwinian Paradigm: Essays on its History, Philosophy, and Religious Implications* (London: Routledge).

—— (1989) *Molecules to Men: The Concept of Progress in Evolutionary Biology* (Cambridge, Mass.: Harvard University Press).

Ruse, M., and E.O. Wilson (1986) Moral philosophy as applied science. *Philosophy,* 61: 173-92.

Russell, E.S. (1916) *Form and Function* (London: Murray).

Russett, C. (1976) *Darwin in America: The Intellectual Response. 1865-1912* (San Francisco: Freeman).

Saarinen, E., ed. (1982) *Conceptual Issues in Ecology* (Dordrecht, Holland: Reidel).

Sahlins, M. (1976) *The Use and Abuse of Biology* (Ann Arbor, Mich.: University of Michigan Press).

Salthe, S.N. (1985) *Evolving Hierarchical Systems: Their Structure and Representation* (New York: Columbia University Press).

Satdinova, N. Kh. (1982) Sotsiobiologiia — "za" i "Protiv". *Voprosy Filosofii*, 3: 129-36.

Sattler, R. (1986) *Biophilosophy: Analytic and Holistic Perspectives* (Berlin: Springer).

Saunders, P.T., and M.-W. Ho (1984) The complexity of organisms. In J. Pollard, ed., *Evolutionary Theory: Paths into the Future* (Chichester, Gt. Britain: J. Wiley and Sons), 121-40.

Savage-Rumbaugh, E.S., and W.D. Hopkins (1986) The evolution of communicative capacities. In W. Bechtel, ed., *Integrating Scientific Disciplines* (Dordrecht, Holland: Nijhoff), 243-62.

Schaffner, K.F. (1967a) Antireductionism and molecular biology. *Science*, 157: 644-47.

—— (1967b) Approaches to reduction. *Philosophy of Science*, 34: 137-47.

—— (1969a) Theories and explanations in biology. *Journal of the History of Biology*, 2: 19-33.

—— (1969b) The Watson-Crick model and reductionism. *British Journal for the Philosophy of Science*, 20: 325-48.

—— (1969c) Chemical systems and chemical evolution: The philosophy of molecular biology. *American Scientist*, 57: 410-20.

—— (1974) The peripherality of reductionism in the development of molecular biology. *Journal of the History of Biology*, 7, (1): 111-39.

—— (1976) Reductionism in biology: Prospects and problems. In R.S. Cohen et al., eds., *PSA 1974* (Dordrecht, Holland: Reidel). 613-32.

—— (1980) Theory structure in the biomedical sciences. *Journal of Medicine and Philosophy*, 5: 57-97.

Schank, J.C., and W.C. Wimsatt (1987) Generative entrenchment and evolution. In A. Fine and P.K. Machamer, eds., *PSA 1986* (East Lansing, Mich.: Philosophy of Science Association), 2: 33-60.

Schneewind, J.B. (1978) Sociobiology, social policy and nirvana. In M.S. Gregory, A. Silvers, and D. Sutch, eds., *Sociobiology and Human Nature* (San Francisco: Jossey-Bass), 225-39.

Scriven, M. (1959) Explanation and prediction in evolutionary theory. *Science,* 130: 477-82.

Searle, J.R. (1978) Sociobiology in the explanation of behaviour. In M.S. Gregory, A. Silvers, and D. Sutch, eds., *Sociobiology and Human Nature* (San Francisco: Jossey-Bass), 164-82.

Segerstrale, U. (1986) Colleagues in conflict: An 'in vivo' analysis of the sociobiology controversy. *Biology and Philosophy,* 1: 53-88.

Selander, R.K. (1982) Phylogeny. In R. Milkman, ed., *Perspectives on Evolution* (Sunderland, Mass.: Sinauer), 32-59.

Shaner, D.E. (1987) The rectification of names. *Biology and Philosophy,* 2: 347-68.

Simberloff, D. (1982) A succession of paradigms in ecology: Essentialism to materialism and probabilism. In E. Saarinen, ed., *Conceptual Issues in Ecology* (Dordrecht, Holland: Reidel), 63-100.

Simpson, G.G. (1961) *Principles of Animal Taxonomy* (New York: Columbia University Press).

Singer, P. (1975) *Animal Liberation* (New York: New York Review).

—— (1981) *The Expanding Circle: Ethics and Sociobiology* (New York: Farrar, Straus, and Giroux).

—— (1986) Life, the universe and ethics: A review of Edward O. Wilson, *Biophilia. Biology and Philosophy,* 1: 367-72.

Sloep, P. (1986) Null hypotheses in ecology: Towards the dissolution of a controversy. In A. Fine and P. Machamer, eds., *PSA 1986* (East Lansing, Mich.: Philosophy of Science Association), 1: 307-13.

Sloep, P., and W. van der Steen (1987a) The nature of evolutionary theory: The semantic challenge. *Biology and Philosophy,* 2: 1-16.

—— (1987b) Syntacticism versus semanticism: Another attempt at dissolution. *Biology and Philosophy,* 2: 33-42.

Smart, J.J.C. (1953) Theory construction. In A.G.N. Flew, ed., *Logic and Language, Second Series* (Oxford: Blackwell), 222-42.

—— (1963) *Philosophy and Scientific Realism* (London: Routledge and Kegan Paul).

Smith, C.U.M. (1987) "Clever beasts who invented knowing": Nietzche's evolutionary biology of knowledge. *Biology and Philosophy*, 2: 65-92.

Sneath, P.H.A., and R.R. Sokal (1973) *Principles of Numerical Taxonomy* 2nd ed. (San Francisco: Freeman).

Snyder, A.A. (1983) Taxonomy and theory. In P. Asquith and T. Nickles, eds., *PSA 1982* (East Lansing, Mich.: Philosophy of Science Association), 2: 512-24.

Sober, E. (1980) Evolution, population thinking and essentialism. *Philosophy of Science*, 47: 350-83.

—— (1981a) The evolution of rationality. *Synthese*, 46: 95-120.

—— (1981b) Revisability, a priori truth, and evolution. *Australasian Journal of Philosophy*, 59: 68-85.

—— (1981c) Holism, individualism, and the units of selection. In P. Asquith and R. Giere, eds., *PSA 1980* (East Lansing, Mich.: Philosophy of Science Association), 2: 93-121.

—— (1983a) The modern synthesis: Its scope and limits. In P. Asquith and T. Nickles, eds., *PSA 1982* (East Lansing, Mich.: Philosophy of Science Association), 2: 314-21.

—— (1983b) Parsimony in systematics: Philosophical issues. *Annual Review of Ecology and Systematics*, 14: 335-57.

—— (1984a) *The Nature of Selection: Evolutionary Theory in Philosophical Focus* (Cambridge, Mass.: The MIT Press).

—— (1984b) Fact, fiction, and fitness. *Journal of Philosophy*, 81: 372-83.

—— (1984c) Discussion: Sets, species and evolution: Comments on Philip Kitcher's 'Species'. *Philosophy of Science*, 51: 334-41.

—— (1984d) Force and disposition in evolutionary theory. In C. Hookway, ed., *Minds, Machines and Evolution* (Cambridge: Cambridge University Press), 43-62.

—— , ed. (1984e) *Conceptual Issues in Evolutionary Biology* (Cambridge, Mass.: The MIT Press).

—— (1985a) Methodological behaviorism, evolution and game theory. In J. Fetzer, ed., *Sociobiology and Epistemology* (Dordrecht, Holland: Reidel), 181-200.

—— (1985b) Panglossian functionalism and the philosophy of mind. *Synthese*, 64: 165-93.

—— (1987a) What is adaptationism? In J. Dupré, ed., *The Latest on the Best* (Cambridge, Mass.: The MIT Press), 105-118.

—— (1987b) Does "fitness" fit the facts? *Journal of Philosophy*, 84: 220-23.

—— (1988a) *Reconstructing the Past: Parsimony, Evolution, and Inference* (Cambridge, Mass.: The MIT Press).

—— (1988b) What is evolutionary altruism? *Canadian Journal of Philosophy*, forthcoming.

Sober, E., and R. Lewontin (1982) Artifact, cause and genic selection. *Philosophy of Science*, 49: 157-62.

—— (1983) Discussion: Reply to Rosenberg on genic selection. *Philosophy of Science*, 50: 648-50.

Sommerhoff, G. (1950) *Analytical Biology* (London: Oxford University Press).

Stanley, S.M. (1979) *Macroevolution: Pattern and Process* (San Francisco: Freeman).

Stanzione, M. (1987) Popper's evolutionary epistemology and analogical reasoning, ms.

Stebbins, G.L. (1977) In defense of evolution: Tautology or theory? *American Naturalist*, 111: 386-90.

—— (1987) Species concepts: Semantics and actual situations. *Biology and Philosophy*, 2: 198-203.

Stebbins, G.L., and F.J. Ayala (1981) Is a new evolutionary synthesis necessary? *Science*, 213: 967-71.

—— (1985) The evolution of Darwinism. *Scientific American*, 253 (1): 72-82.

Steele, E.J., et al., (1984) The somatic selection of acquired characteristics. In J. Pollard, ed., *Evolutionary Theory: Paths into the Future* (Chichester, Gt. Britain: J. Wiley and Sons), 217-38.

Stenseth, N.C., A. Tjonneland, and T. Lindholm (1988) Can rationality and irrationality be reconciled? *Biology and Philosophy*, 3: forthcoming.

Stent, G.S. (1985) Hermeneutics and the analysis of complex biological systems. In D.J. Depew and B.H. Weber, eds., *Evolution at a Crossroads* (Cambridge, Mass.: The MIT Press), 209-26.

—— (1986) Glass bead game: A review of Alexander Rosenberg, *The Structure of Biological Science. Biology and Philosophy*, 1: 227-48.

Stidd, B.M. (1980) The neotenous origin of the pollen organ of the gymnosperm *Cycadeoidea* and the implications for the origin of higher taxa. *Paleobiology*, 6: 161-67.

Stillings, N. et al., (1987) *Cognitive Science* (Cambridge, Mass.: The MIT Press).

Stich S. (1978) The recombinant DNA debate. *Philosophy and Public Affairs*, 7: 187-205.

Strong, D.R. (1982) Null hypotheses in ecology. In E. Saarinen, ed., *Conceptual Issues in Ecology* (Dordrecht, Holland: Reidel), 245-60.

Suppe, F. (1977) *The Structure of Scientific Theories.* rev. ed. (Urbana: University of Illinois Press).

Symons, D. (1979) *The Evolution of Human Sexuality* (New York: Oxford University Press).

Taylor, C. (1964) *The Explanation of Behaviour* (London: Routledge and Kegan Paul).

Taylor, P. (1986) *Respect for Nature: A Theory of Environmental Ethics* (Princeton, N.J.: Princeton University Press).

Templeton, A.R. (1982) Adaptation and the integration of evolutionary forces. In R. Milkman, ed., *Perspectives on Evolution* (Sunderland, Mass.: Sinauer), 15-31.

Tennant, N. (1983a) Evolutionary vs evolved ethics. *Philosophy*, 58: 289-302.

—— (1983b) Evolutionary epistemology. In *Epistemology and Philosophy of Science: Proceedings of the 7th International Wittgenstein Symposium* (Vienna: Holder-Pichler-Tempsky), 168-73.

—— (1983c) In defence of evolutionary epistemology. *Theoria*, 49: 32-48.

—— (1984) Intentionality, syntactic structure and the evolution of language. In C. Hookway, ed., *Minds, Machines and Evolution* (Cambridge: Cambridge University Press), 73-104.

—— (1985) Beth's Theorem and reductionism. *Pacific Philosophical Quarterly*, 66: 342-54.

—— (1986) Reductionism and holism in biology. In T.J. Horder, J.A. Witkowsky, and C.C. Wylie, eds., *A History of Embryology* (Cambridge: Cambridge University Press), 407-33.

—— (1987) Philosophy and biology: mutual enrichment or one-sided encroachment? *La Nuova Critica*, 1/2 n.s.: 39-53.

—— (1988a) Theories, concepts and rationality in an evolutionary account of science. *Biology and Philosophy*, 3: 224-31.

—— (1988b) Two problems for evolutionary epistemology: psychic reality and the emergence of norms. *Ratio*, forthcoming.

Thagard, P. (1980) Against evolutionary epistemology. In P. Asquith and R. Giere, eds., *PSA 1980* (East Lansing, Mich.: Philosophy of Science Association), 1: 187-96.

Thomas, L. (1986) Biological moralism. *Biology and Philosophy*, 1: 316-24.

Thompson, R.P. (1980) Is sociobiology a pseudoscience? In P. Asquith and R. Giere, eds., *PSA 1980* (East Lansing, Mich.: Philosophy of Science Association), 1: 363-70.

—— (1983a) Tempo and mode in evolution. Punctuated equilibria and the modern synthetic theory. *Philosophy of Science*, 50: 432-52.

—— (1983b) The structure of evolutionary theory: A semantic approach. *Studies in History and Philosophy of Science*, 14: 215-29.

—— (1983c) Historical laws in modern biology. *Acta Biotheoretica*, 32: 167-77.

—— (1985) Sociobiological explanation and the testability of sociobiological theory. In J. Fetzer, ed., *Sociobiology and Epistemology* (Dordrecht, Holland: Reidel), 201-15.

—— (1986) The interaction of theories and the semantic conception of evolutionary theory. *Philosophica*, 37: 73-86.

—— (1987) A defense of the semantic conception of evolutionary theory. *Biology and Philosophy*, 2: 26-32.

—— (1988a) *The Structure of Biological Theories* (Albany, N.Y.: State University of New York Press).

—— (1988b) Logical and epistemological aspects of the 'new' evolutionary epistemology. *Canadian Journal of Philosophy*, forthcoming.

—— (1988c) The conceptual role of intelligence in human sociobiology. In H.J. Jerison and I. Jerison, eds., *Intelligence and Evolutionary Biology* (New York: Springer-Verlag), 35-44.

—— (1988d) David Hull's conception of the structure of evolutionary theory. In M. Ruse, ed., *What the Philosophy of Biology Is* (Dordrecht, Holland: Kluwer).

—— (1988e) Some punctuationists *are* wrong about the modern synthesis. *Philosophy of Science* 55, 74-86.

—— (1988f) Philosophy of biology under attack — Stent vs. Rosenberg. *Biology and Philosophy*, 3, forthcoming.

Toulmin, S. (1967) The evolutionary development of natural science. *American Scientist*, 57: 456-71.

—— (1972) *Human Understanding* (Oxford: Clarendon Press).

Trigg, R. (1982) *The Shaping of Man* (Oxford: Blackwell).

—— (1986) Evolutionary ethics. *Biology and Philosophy*, 1: 325-36.

Trivers, R.L. (1972) Parental investment and sexual selection. In B. Campbell, ed., *Sexual Selection and the Descent of Man, 1871-1971* (Chicago: Aldine).

—— (1976) "Foreword" to R. Dawkins, *The Selfish Gene* (Oxford: Oxford University Press), v-vii.

Turner, J.R.G. (1983) "The hypothesis that explains mimetic resemblance explains evolution": The gradualist-saltationist schism. In M. Grene, ed., *Dimensions of Darwinism* (Cambridge: Cambridge University Press), 129-69.

Ulanowicz, R.E. (1986) *Growth and Development* (New York: Springer-Verlag).

Utz, S. (1977) Discussion: On teleology and organisms. *Philosophy of Science*, 44: 313-20.

Van Balen, G. (1987) Conceptual tensions between theory and program: The chromosome theory and the Mendelian research program. *Biology and Philosophy*, 2: 435-62.

Van den Berghe, P. (1979) *Human Family Systems* (New York: Elsevier).

Van der Steen, W.J. (1983a) Methodological problems in evolutionary biology. 1. Testability and tautologies. *Acta Biotheoretica*, 32: 207-15.

—— (1983b) Methodological problems in evolutionary biology. 2. Appraisal of arguments against adaptationism. *Acta Biotheoretica*, 32: 217-22.

—— (1986a) Methodological problems in evolutionary biology. The force of evolutionary epistemology. *Acta Biotheoretica*, 35: 193-204.

—— (1986b) Methodological problems in evolutionary biology. The import of supervenience. *Acta Biotheoretica*, 35: 185-91.

Van der Steen, W.J., and B. Voorzanger (1984a) Methodological problems in evolutionary biology. 3. Selection and levels of organization. *Acta Biotheoretica*, 33: 199-213.

—— (1984b) Sociobiology in perspective. *Journal of Human Evolution*, 13: 25-32.

—— (1986) Methodological problems in evolutionary biology. The species plague. *Acta Biotheoretica*, 35: 205-21.

Van Fraassen, B.C. (1980) *The Scientific Image* (Oxford: Oxford University Press).

Vollmer, G. (1975) *Evolutionäre Erkenntnistheorie* (Stuttgart: Hirzel).

Von Schilcher, F., and N. Tennant (1984) *Philosophy, Evolution and Human Nature* (London: Routledge and Kegan Paul).

Voorzanger, B. (1984) Altruism in sociobiology: A conceptual analysis. *Journal of Human Evolution*, 13: 33-39.

—— (1987a) No norms and no nature — the moral relevance of evolutionary biology. *Biology and Philosophy*, 2: 253-70.

—— (1987b) Methodological problems in evolutionary biology. Biology and culture. *Acta Biotheoretica*, 36: 23-34.

Wachbroit, R. (1986) Progress: metaphysical and otherwise. *Philosophy of Science*, 53: 354-71.

Waddington, C.H. (1957) *The Strategy of the Genes* (London: Allen and Unwin).

—— (1960) *The Ethical Animal* (London: Allen and Unwin).

—— , ed. (1968) *Towards a Theoretical Biology, 1, Prolegomena* (Edinburgh: Edinburgh University Press).

—— , ed. (1969) *Towards a Theoretical Biology, 2, Sketches* (Edinburgh: Edinburgh University Press).

—— , ed. (1970) *Towards a Theoretical Biology, 3, Drafts* (Edinburgh: Edinburgh University Press).

Wade, M.J. (1978) A critical review of the models of group selection. *Quarterly Review of Biology*, 53: 101-14.

Washburn, S.L. (1978) Animal behaviour and social anthropology. In M.S. Gregory, A. Silvers, and D. Sutch, eds., *Sociobiology and Human Nature* (San Francisco: Jossey-Bass), 53-74.

Wassermann, G. (1978) The testability of the role of natural selection within theories of population genetics and evolution. *British Journal for the Philosophy of Science*, 29: 223-42.

—— (1981) On the nature of the theory of evolution. *Philosophy of Science*, 48: 416-37.

Waters, C.K. (1986a) Natural selection without survival of the fittest. *Biology and Philosophy*, 1: 202-26.

—— (1986b) Taking analogical inference seriously: Darwin's argument from artificial selection. In A. Fine and P. Kitcher, eds., *PSA 1986* (East Lansing, Mich: Philosophy of Science Association), 1: 502-13.

Watson, J.D. (1968) *The Double Helix* (New York: Atheneum).

Watson, J.D., and F.H.C. Crick (1953) Molecular structure of nucleic acids, *Nature*, 171: 737-38.

Weber, B.H., D.J. Depew, and J.D. Smith, eds. (1988) *Entropy, Information, and Evolution. New Perspectives on Physical and Biological Evolution* (Cambridge, Mass.: The MIT Press).

Webster, G., and B.C. Goodwin (1982) The origin of species: A structuralist approach. *Journal of Social and Biological Structures*, 5: 15-47.

Whewell, W. (1840) *Philosophy of the Inductive Sciences* (London: Parker).

Wicken, J. (1986) Evolutionary self-organization and entropic dissipation in biological and socioeconomic systems. *Journal of Social and Biological Structures*, 9: 261-73.

Wiley, E.O. (1975) Karl R. Popper, systematics, and classification: A reply to Walter Bock and other evolutionary taxonomists. *Systematic Zoology*, 24: 233-43.

—— (1981) *Phylogenetics: The Theory and Practice of Phylogenetic Systematics* (New York: J. Wiley and Sons).

Wiley, E.O. and D.R. Brooks (1982) Victims of history — a nonequilibrium approach to evolution. *Systematic Zoology*, 31: 1-24.

—— (1987) A response to Professor Morowitz. *Biology and Philosophy*, 2: 369-74.

Williams, G.C. (1966) *Adaptation and Natural Selection: A Critique of Some Current Evolutionary Thought* (Princeton, N.J.: Princeton University Press).

—— (1984) Group selection. In R.N. Brandon and R.M. Burian, eds., *Genes, Organisms and Populations* (Cambridge, Mass.: The MIT Press), 52-68.

—— (1985a) A defense of reductionism in evolutionary biology. In R. Dawkins and M. Ridley, eds., *Oxford Surveys in Evolutionary Biology* (Oxford: Oxford University Press), 2: 1-27.

—— (1985b) Discussion: Elliott Sober's *The Nature of Selection. Biology and Philosophy*, 1: 109-24.

Williams, M.B. (1970) Deducing the consequences of evolution: A mathematical model. *Journal of Theoretical Biology*, 29: 343-85.

—— (1973) Falsifiable predictions of evolutionary theory. *Philosophy of Science*, 40: 518-37.

—— (1976) The logical structure of functional explanations in biology. In F. Suppe and P. Asquith, eds., *PSA 1986* (East Lansing, Mich.: Philosophy of Science Association), 1: 37-46.

—— (1981) Similarities and differences between evolutionary theory and the theories of physics. In P. Asquith and R. Giere, eds., *PSA 1980* (East Lansing, Mich.: Philosophy of Science Association), 2: 385-96.

—— (1985) Species are individuals: Theoretical foundations for the claim. *Philosophy of Science*, 52: 578-90.

—— (1986) The logical skeleton of Darwin's historical methodology. In A. Fine and P. Kitcher, eds., *PSA 1986* (East Lansing, Mich.: Philosophy of Science Association), 1: 514-21.

—— (1987) A criterion relating singularity and individuality. *Biology and Philosophy*, 2: 204-6.

—— (1988) Evolvers are individuals: Extension of the species as individuals claim. In M. Ruse, ed., *What the Philosophy of Biology Is* (Dordrecht, Holland: Kluwer).

Wilson, D.S. (1980) *The Natural Selection of Populations and Communities* (Menlo Park, Calif.: Benjamin Cummings).

—— (1983) The group selection controversy: History and current status. *Annual Review of Ecology and Systematics*, 14: 159-88.

—— (1984) Individual selection and the concept of structured demes. In R.N. Brandon and R.M. Burian, eds., *Genes, Organisms and Populations* (Cambridge, Mass.: The MIT Press), 272-91.

Wilson, E.O. (1971) *The Insect Societies* (Cambridge, Mass.: Harvard University Press).

—— (1975) *Sociobiology: The New Synthesis* (Cambridge, Mass.: Harvard University Press).

—— (1978) *On Human Nature* (Cambridge, Mass.: Harvard University Press).

—— (1984) *Biolphilia* (Cambridge, Mass.: Harvard University Press).

—— (1985) The sociogenesis of insect colonies. *Science*, 228: 1489-95.

Wimsatt, W.C. (1971) Some problems with the concept of feedback. In R.C. Buck and R.S. Cohen, eds., *PSA 1970* (Dordrecht, Holland: Reidel), 241-56.

—— (1972) Teleology and the logical structure of function statements. *Studies in History and Philosophy of Science*, 3: 1-80.

—— (1974) Complexity and organization. In K.F. Schaffner and R.S. Cohen, eds., *PSA 1972* (Dordrecht, Holland: Reidel), 67-86.

—— (1976a) Reductionism, levels of organization and the mind-body problem. In G. Globus, I. Savodnik, and G. Maxwell, eds., *Consciousness and the Brain* (New York: Plenum), 199-267.

—— (1976b) Reductive explanation: A functional account. In R.S. Cohen et al., eds., *PSA 1974* (Dordrecht, Holland: Reidel), 671-710.

—— (1979) Reduction and reductionism. In P.D. Asquith and H. Kyburg, Jr., eds., *Current Research in Philosophy of Science* (East Lansing, Mich.: Philosophy of Science Association), 352-77.

—— (1980a) Reductionistic research strategies and their biases in the units of selection controversy. In T. Nickles, ed., *Scientific Discoveries: Case Studies* (Dordrecht, Holland: Reidel), 213-59.

—— (1980b) Randomness and perceived randomness in evolutionary biology. *Synthese*, 43: 287-329.

—— (1981a) The units of selection and the structure of the multilevel genome. In P. Asquith and R. Giere, eds., *PSA 1980* (East Lansing, Mich.: Philosophy of Science Association), 2: 122-83.

—— (1981b) Robustness, reliability and overdetermination. In M. Brewer and B. Collins, eds., *Scientific Inquiry and the Social Sciences* (San Francisco: Jossey-Bass), 124-63.

—— (1985a) Heuristics and the study of human behaviour. In D.W. Fiske and R. Shweder, eds., *Metatheory in Social Science: Pluralisms and Subjectivities* (Chicago: University of Chicago Press), 293-314.

—— (1985b) Forms of aggregativity. In A. Donagan, N. Perovich, and M. Wedin, eds., *Human Nature and Natural Knowledge* (Dordrecht, Holland: Reidel), 259-93.

—— (1986) Developmental constraints, generative entrenchment and the innate-acquired distinction. In W. Bechtel, ed., *Integrating Scientific Disciplines* (Dordrecht, Holland: Nijhoff), 185-208.

—— (1987) False models as means to truer theories. In N. Nitecki and A. Hoffman, eds., *Neutral Models in Biology* (Oxford: Oxford University Press), 23-55.

Wimsatt, W.C., and J.C. Schank (1988) Two constraints on the evolution of complex adaptations and the means for their avoidance. In M. Nitecki, ed., *The Idea of Progress in Evolution* (Chicago: University of Chicago Press).

Woodfield, A. (1976) *Teleology* (Cambridge: Cambridge University Press).

Woodger, J.H. (1937) *The Axiomatic Method in Biology* (Cambridge: Cambridge University Press).

—— (1939) *The Technique of Theory Construction* (Chicago: University of Chicago Press).

—— (1952) *Biology and Language* (Cambridge: Cambridge University Press).

—— (1959) Studies in the foundations of genetics. In L. Henkin, P. Suppes, and A. Tarski, eds., *The Axiomatic Method* (Amsterdam: North-Holland).

Wright, L. (1972) Discussion: A comment on Ruse's analysis of function statements. *Philosophy of Science*, 39: 512-14.

—— (1973) Functions, *Philosophical Review*, 82: 139-68.

—— (1974) Mechanisms and purposive behaviour III. *Philosophy of Science*, 41: 345-60.

—— (1976) *Teleological Explanations* (Berkeley, Calif.: University of California Press).

Wuketits, F.M. (1978) *Wissenschafts Theoretische Probleme der Modernen Biologie* (Berlin: Duncker and Humbolt).

——, ed. (1983) *Concepts and Approaches in Evolutionary Epistemology* (Dordrecht, Holland: Reidel).

—— (1986) Evolution as a cognition process: Towards an evolutionary epistemology. *Biology and Philosophy*, 1: 191-202.

Young, J.Z. (1987) *Philosophy and the Brain* (Oxford: Oxford University Press).

Zanardo, A. and Rizzotti, M. (1986) Axiomatization of genetics 2. Formal development. *Journal of Theoretical Biology*, 118: 145-52.

Index

DATE DUE

GAYLORD			PRINTED IN U.S.A.